Additive groups of rings

S Feigelstock

Bar-Ilan University

Additive groups of rings

Pitman Advanced Publishing Program

BOSTON · LONDON · MELBOURNE

PITMAN BOOKS LIMITED
128 Long Acre, London WC2E 9AN

PITMAN PUBLISHING INC
1020 Plain Street, Marshfield, Massachusetts 02050

Associated Companies
Pitman Publishing Pty Ltd, Melbourne
Pitman Publishing New Zealand Ltd, Wellington
Copp Clark Pitman, Toronto

First published 1983

AMS Subject Classifications: (main) 20K99
(subsidiary) 16, 17

Library of Congress Cataloging in Publication Data

Feigelstock, S.
Additive groups of rings.

(Research notes in mathematics; 83)
Bibliography: p.
Includes index.
1. Abelian groups. I. Title. II. Series.
QA171.F34 1983 512′.2 82-22412
ISBN 0-273-08591-3

British Library Cataloguing in Publication Data

Feigelstock, S.
Additive groups of rings.—(Research notes in mathematics; 83)
1. Group rings
I. Title II. Series
512′.4 QA171

ISBN 0-273-08591-3

Reproduced and printed by photolithography
in Great Britain by Biddles Ltd, Guildford

Preface

In 1948 R.A. Beaumont [5] began investigating the additive groups of rings. Shortly thereafter L. Redei, T. Szele [54], and H.S. Zuckerman [11] joined in these investigations, to be followed by L. Fuchs [37], [38]. Since then much progress has been made in this branch of abelian group theory. Many of the results appear in Chapter 17 of L. Fuchs' well-known book [36]. An attempt has been made here to offer an easily accessible account of much of the work done on the additive groups of rings, with special emphasis on results not covered in [36]. This short monograph is far from being comprehensive, and many important papers have not been considered here at all.

Although this work is meant to be self contained, some knowledge of abelian group theory is assumed. The prospective reader is especially urged to read Chapter 17 of Fuchs' book [36].

I wish to express my gratitude to Professor Elvira Rapaport Strasser for introducing me to this subject, for her advice and guidance during the past fifteen years, and for suggesting that I write this monograph. I also thank Professor L. Fuchs for his valuable comments and suggestions.

Contents

Notation

G	a group.
$\lvert x\rvert$	the order of $x \in G$.
G_t	the torsion part of G.
G_p	the p-primary component of G, p a prime.
nG	$\{nx \mid x \in G\}$, n a fixed integer.
$G[n]$	$\{x \in G \mid nx = 0\}$, n a fixed integer.
G^α	the α-th Ulm subgroup of G, α an ordinal.
$\nu(G)$	the nilstufe of G.
$N(G)$	the strong nilstufe of G.
$r(G)$	the rank of G.
(S)	the subgroup of G generalized by $S \subseteq G$.
$h(x)$	the height of $x \in G$.
$h_p(x)$	the p-component of $h(x)$.
$t(x)$	the type of $x \in G$.
$t(G)$	the type of G for G homogeneous.
$T(G)$	$\{t(x) \mid x \in G,\ x \neq 0\}$ = the type set of G.
$t_p(x)$ $(t_p(G))$	the p-component of $t(x)$ $(t(G))$.
$[(k_1, k_2, \ldots, k_n, \ldots)]$	the type containing the height vector $(k_1, k_2, \ldots, k_n, \ldots)$. Additional notation concerning type which will be used may be found in [36, vol. 2, pp. 108-111].
$Z(n)$	cyclic group of order n.
$Z(p^\infty)$	p-Prüfer group.
$\mathrm{End}(G)$	the group of endomorphisms of G.
$\dot{\mathrm{E}}\mathrm{nd}(G)$	the group of quasi-endomorphisms of G.
$\dot{\cong}$	quasi-isomorphism.
$\doteq$	quasi-equality
R	a ring.
R^+	the additive group of R.

$I \triangleleft R$	"I is an ideal in R".
$\langle S \rangle$	the ideal in R generated by $S \subseteq R$
$M_n(R)$	the ring of $n \times n$ matrices with components in R, n a positive integer.
Z	the ring of integers.
Q	the field of rational numbers.
Z_p	the ring of p-adic integers.
ω	the first infinite ordinal
$\|S\|$	the cardinality of a set S.

Additional symbols which appear in only one section, will be defined in the section they appear in.

1 Preliminaries

§1. DEFINITIONS AND STRUCTURE THEOREMS

All groups considered in this monograph are abelian, with addition the group operation.

Definition: A group G is p-divisible, p a prime, if $G = pG$. If G is p-divisible for every prime p, then G is said to be divisible.

It is readily seen that G is divisible if and only if $G = nG$ for every positive integer n.

Definition: A subgroup H of a group G is p-pure in G, p a prime, if $H \cap p^kG = p^kH$ for every positive integer k. If H is p-pure in G for every prime p, then H is said to be pure in G. Clearly H is pure in G if and only if $H \cap nG = nH$ for every integer n.

Definition: A subset $\{a_i \mid i \in I\}$ of a group G is said to be independent if for distinct $i_1,\ldots,i_k \in I$, k an arbitrary positive integer, and for $n_1,\ldots,n_k \in Z$, $n_1a_{i_1}+\ldots+n_ka_{i_k} = 0$ implies that $n_1a_{i_1} = \ldots = n_ka_{i_k} = 0$.

By Zorn's lemma, there exists a maximal independent set M in G containing only elements of infinite and prime power orders. The rank of G, $r(G) = |M|$. Let M_0 be maximal amongst the independent subsets of G containing only torsion free elements. Then $|M_0|$ is called the torsion free rank of G, denoted $r_0(G)$. Let M_p be maximal amongst the independent subsets of G containing only elements of order a power of p, p a prime. Then $|M_p|$ is called the p-rank of G, denoted $r_p(G)$.

$r(G)$, $r_0(G)$, and $r_p(G)$ are well defined [36, Theorem 16.3].

Definition: A subgroup B of a group G is a p-basic subgroup, p a prime if:

1) B is a direct sum of infinite cyclic groups, and cyclic p-groups.
2) B is p-pure in G, and
3) G/B is p-divisible.

If G is a p-group, then a p-basic subgroup of G is called a basic subgroup.

Every group G possesses a p-basic subgroup [36, Theorem 32.3], and all p-basic subgroups of G are isomorphic [36, Theorem 35.2].

<u>Definition</u>: For every ordinal α, define the α-th Ulm subgroup G^{α} of G as follows: $G^1 = \bigcap_{n<\omega} nG$. For every ordinal α, the α+1th Ulm subgroup of G, $G^{\alpha+1} = (G^{\alpha})^1$. For β a limit ordinal, the βth Ulm subgroup of G, $G^{\beta} = \bigcap_{\gamma<\beta} G^{\gamma}$.

<u>Definition:</u> Let G be a group, and let $x \in G$. The p-height of x, $h_p(x)$, is the greatest positive integer n, if it exists, such that $x \in p^nG$. If no such positive integer exists, then $h_p(x) = \infty$. Let $\{p_i\}_{i=1}^{\infty}$ be the sequence of all primes in ascending order. The height (characteristic) of x, $h(x) = (h_{p_1}(x),\ldots,h_{p_n}(x),\ldots)$.

A sequence $(k_1,\ldots,k_n,\ldots)$ with k_n a non-negative integer or ∞ for every positive integer n, is called a height sequence. The relation on the set of height sequences defined by $(k_1,\ldots,k_n,\ldots) \sim (\ell_1,\ldots,\ell_n \ldots)$ if $k_n = \ell_n$ for all but finitely many subscripts n, and $k_n = \ell_n$ whenever $k_n = \infty$ or $\ell_n = \infty$, is an equivalence relation. The equivalence class of a height sequence $(k_1,\ldots,k_n,\ldots)$ is called a type, and will be denoted $[(k_1,\ldots,k_n,\ldots)]$. The type of $x \in G$, $t(x) = [h(x)]$.

<u>Definition</u>: The type set of a group G, $T(G) = \{t(x) | x \in G, x \neq 0\}$. If $|T(G)| = 1$, then G is said to be a homogeneous group with type $t(G)$ = the singleton belonging to $T(G)$.

<u>Definition</u>: A height sequence $(k_1,\ldots,k_n,\ldots)$ is idempotent, if $k_n = 0$ or ∞ for every positive integer n. A type τ is idempotent if τ possesses an idempotent height sequence.

<u>Definition</u>: A group G is indecomposable if $G = H \oplus K$ implies that $H = 0$ or $K = 0$.

<u>Definition</u>: A torsion free group G is rigid, if $End(G)$ is isomorphic to a subgroup of Q^+.

Definition: Let G and H be subgroups of a divisible torsion free group D, in lieu of Proposition 1.1.6, this is not a very severe restriction on torsion free groups G, H. If there exists a positive integer n such that $nG \subseteq H$, and $nH \subseteq G$, then G and H are said to be quasi-equal, $G \doteq H$. If two groups A and B are isomorphic to quasi-equal groups, G, H respectively, then A and B are said to be quasi-isomorphic, $A \doteqdot B$.

Definition: A group G satisfying, $G \doteqdot H \oplus K$ implies $H = 0$ or $K = 0$, is said to be strongly indecomposable.

Definition: A subgroup H of a group G is essential in G if $H \cap K \neq 0$ for every subgroup $K < G$, $K \neq 0$.

Definition: Let G be a torsion free group, and let D be a divisible torsion free group such that G is an essential subgroup of D; such a D exists, Proposition 1.1.6. An endomorphism $\varphi \in \mathrm{End}(D)$ is a quasi-endomorphism of G, if there exists a positive integer n such that $n\varphi \in \mathrm{End}(G)$.

Definition: A group G is bounded if there exists a positive integer n such that $nx = 0$ for all $x \in G$.

Definition: A subgroup H of a group G is fully invariant in G if $\varphi(H) \subseteq H$ for every endomorphism φ of G.

Several structure theorems on abelian groups play a vital role in studying the additive groups of rings. These results will be stated here without proof. However, the reader will be referred to the proofs in Fuchs' book [36].

Proposition 1.1.1: Let G be a torsion group. Then $G = \bigoplus_{p \text{ a prime}} G_p$.

Proof: [36, Theorem 8.4].

Proposition 1.1.2: Let T be a torsion group. T is a direct summand of every group G with $G_t = T$ if and only if T is the direct sum of a divisible group, and a bounded group.

Proof: [36, Theorem 100.1]

Proposition 1.1.3: A group D is divisible if and only if $D \simeq \bigoplus_{\alpha} Q^+ \oplus \bigoplus_{p \text{ a prime}} \bigoplus_{\alpha_p} Z(p^\infty)$, α, α_p, arbitrary cardinal numbers.

Proof: [36, Theorem 23.1].

Proposition 1.1.4. Let G be a non-torsion free group. Then $G = Z(p^k) \oplus H$, for some prime p, $1 \leq k \leq \infty$.

Proof: [36, Corollary 27.3].

Corollary 1.1.5: Let G be a non-torsion free group. G is indecomposable if and only if $G = Z(p^k)$, p a prime, $1 \leq k \leq \infty$.

Proposition 1.1.6: Every group G can be embedded, as an essential subgroup, into a divisible group D satisfying $r_0(D) = r_0(G)$, and $r_p(D) = r_p(G)$, for every prime p.

Proof: [36, Theorem 24.1, Lemma 24.3, and Theorem 24.4].

The group D in Proposition 1.1.6 is unique up to isomorphism, and is called the divisible hull of G, [36, Theorem 24.4].

Proposition 1.1.7: Let G be a rank one torsion free group. Then G is homogeneous and isomorphic to a subgroup of Q^+. Two rank one torsion free groups are isomorphic if and only if they have the same type. For every type τ, there exists a rank one torsion free group G with $t(G) = \tau$.

Proof: [36, Theorem 85.1], Proposition 1.1.6 and Proposition 1.1.3.

Proposition 1.1.8: Quasi-isomorphic rank one torsion free groups are isomorphic.

Proof: [36, remark preceding Proposition 92.1].

Proposition 1.1.9: A bounded group is a direct sum of cyclic groups.

Proof: [36, Theorem 17.2].

Proposition 1.1.10: A divisible group is a direct summand of every group containing it.

Proof: [36, Theorem 21.2].

§2. MULT G

Let G be a group, and let $\times$ be a binary operation on G inducing a ring structure on G. Clearly $\times$ may be viewed as a bilinear map from the cartesian product $G \times G$ to G via $(g_1, g_2) \to g_1 \times g_2$ for all $g_1, g_2 \in G$. Hence there exists $\varphi \in \mathrm{Hom}(G \otimes G, G)$ such that $g_1 \times g_2 = \varphi(g_1 \otimes g_2)$ for all $g_1, g_2 \in G$. Conversely for $\varphi \in \mathrm{Hom}(G \otimes G, G)$ the multiplication

$g_1 \times_\varphi g_2 = \varphi(g_1 \otimes g_2)$ for $g_1, g_2 \in G$ induces a ring structure on G. Therefore the ring multiplication on G form a group Mult G. Addition in Mult G is as follows: Let $x_1, x_2 \in$ Mult G, and let $g_1, g_2 \in G$. Then $g_1(x_1 + x_2)g_2 = (g_1 x_1 g_2) + (g_1 x_2 g_2)$.

1.2.1. Mult $G \simeq \text{Hom}(G \otimes G, G) \simeq \text{Hom}(G, \text{End}(G))$.

The second isomorphism follows from the well-known fact that the functor $\otimes$ is left adjoint to the functor Hom.

The above facts as well as additional information concerning Mult G may be found in [36, section 118].

§3. TYPE

In this section all groups are assumed to be torsion free.

Definition: Let τ_1, τ_2 be types. Then $\tau_1 \geq \tau_2$ means that there exist height vectors $(k_1, \ldots, k_n, \ldots)$, $(\ell_1, \ldots, \ell_n, \ldots)$ belonging to τ_1 and τ_2 respectively, such that $k_n \geq \ell_n$; $n = 1, 2, \ldots$.

Definition: Let $\tau_1 = [(k_1, \ldots, k_n, \ldots)]$, $\tau_2 = [(\ell_1, \ldots, \ell_n, \ldots)]$ be two types. The product of τ_1 and τ_2 is defined to be the type $\tau_1 \cdot \tau_2 = [(k_1 + \ell_1, \ldots, k_n + \ell_n, \ldots)]$. The intersection of τ_1 and τ_2, $\tau_1 \cap \tau_2 = [(\min(k_1, \ell_1), \ldots, \min(k_n, \ell_n), \ldots)]$. If $k_n \geq \ell_n$ for every positive integer n, then the quotient $\tau_1 : \tau_2 = [(k_1 - \ell_1, \ldots, k_n - \ell_n, \ldots)]$ with the convention $\infty - k = \infty$ for $0 \leq k \leq \infty$.

Property 1.3.1: 1) Let $g, h \in G$ be dependent. Then $t(g) = t(h)$. 2) $t(g+h) \geq t(g) \cap t(h)$ for all $g, h \in G$. 3) For $G = H \oplus K$, $h \in H$, $k \in K$, $t(h+k) = t(h) \cap t(k)$. 4) For $\varphi: G \to H$ a group homomorphism, and $g \in G$, $t[\varphi(g)] \geq t(g)$.

Proof: [36, p. 109].

Consequence 1.3.2: Let R be a ring, $a,b \in R$. Then $t(ab) \geq \max[t(a), t(b)]$.

Proof: Property 1.3.1.4 and the fact that the maps $x \to ax$, $x \to xb$ are endomorphisms of R^+.

Proposition 1.3.3: Let G,H be rank one torsion free groups. Then $G \otimes H$ is a rank one torsion free group with $t(G \otimes H) = t(G) \cdot t(H)$.

Proof: [36, Proposition 85.3].

Proposition 1.3.4: Let G,H be rank one torsion free groups. If $t(G) \leq t(H)$ then $Hom(G,H) = 0$. If $t(G) \leq t(H)$, then $Hom(G,H)$ is a rank one torsion free group with $t[Hom(G,H)] = t(H): t(G)$.

Proof: [36, Proposition 85.4].

§4: Examples:

1.4.1. $F = \bigoplus_{\alpha} Z^{+}$, α an arbitrary cardinal.

Let $\{a_i \mid i \in I\}$ be a basis for F. By 1.2.1, every $\varphi \in Hom(F \otimes F, F)$ determines a ring multiplication on F. Since $F \otimes F$ is a free abelian group with basis $\{a_i \otimes a_j \mid i,j \in I\}$, any map of these basis elements into F may be linearly extended to a homomorphism mapping $F \otimes F$ into F. Therefore the rings with additive group F are determined by arbitrarily defining the products $a_i \cdot a_j$, $i,j \in I$, and extending the multiplication linearly.

1.4.2. $G = \bigoplus_{\alpha} Q^{+}$, α an arbitrary cardinal.

Let $\{a_i \mid i \in I\}$ be a basis for G, viewed as a vector space over Q. Then $\{a_i \otimes a_j \mid i,j \in I\}$ is a basis over Q for $G \otimes G$. The same argument used in 1.4.1, with the exception that $G \otimes G$ is a free object in the category of vector spaces over Q, while $F \otimes F$ is free in the category of abelian groups, shows that the rings with additive group G are determined by arbitrarily defining $a_i \cdot a_j$, $i,j \in I$ and extending linearly.

1.4.3. $G = \bigoplus_{\alpha} Z(p)$, p a prime, α an arbitrary cardinal.

The same argument employed in 1.4.2, except that now G is viewed as a vector space over a field of order p, shows that the rings with additive group G are obtained by arbitrarily defining the products of basis elements, and extending linearly.

1.4.4. $G = \bigoplus_{j<\omega} \bigoplus_{\alpha_j} Z(p^j)$, p a prime, α_j an arbitrary cardinal, $j = 1, 2, \ldots$, i.e., G is any direct sum of cyclic p-groups.

Let $\{a_{ij} \mid i \in I_j\}$ be a basis for $\bigoplus_{\alpha_j} Z(p^j)$, $j = 1, 2, \ldots$ Then $\{a_{i_1 j_1} \otimes a_{i_2 j_2} \mid i_1 \in I_{j_1},\ i_2 \in I_{j_2},\ j_1, j_2 = 1, 2, \ldots\}$ is a basis for $G \otimes G$, with $|a_{i_1j_1} \otimes a_{i_2j_2}| = p^{\min(j_1,j_2)}$. The homomorphisms $\varphi \in \mathrm{Hom}(G\otimes G, G)$ are determined by defining $\varphi(a_{i_1j_1} \otimes a_{i_2j_2})$ to be any element in G of order less than or equal $p^{\min(j_1,j_2)}$. Therefore the ring multiplication on G are determined by defining the product of basis elements with the sole restriction $|ab| \leq \min(|a|, |b|)$ for every pair of basis elements a,b.

In all of the above examples, the ring multiplications which were defined are associative (commutative) if the product of basis elements is associative (commutative).

1.4.5. G a torsion group.

$G = \bigoplus_{p \text{ a prime}} G_p$, Proposition 1.1.1. It is readily seen that $\mathrm{Hom}(G\otimes G, G) \simeq \prod_{p \text{ a prime}} \mathrm{Hom}(G_p \otimes G_p, G_p)$ i.e., every ring R with $R^+ = G$ is a ring direct sum $R = \bigoplus_{p \text{ a prime}} R_p$ with $R_p^+ = G_p$. Therefore the investigation of the ring structures which may be defined on a torsion group G, reduces to the p-primary case, p a prime.

1.4.6. G a p-primary group, p a prime.

Let B be a basic subgroup of G with basis $\{a_i | i \in I\}$. Then $G\otimes G = B\otimes B$, [36, Theorem 61.1] and so $\mathrm{Hom}(G\otimes G, G) \simeq \mathrm{Hom}(B\otimes B, G)$. As in example 1.4.4, the homomorphisms φ from $B\otimes B$ into G are determined by defining $\varphi(a_i \otimes a_j)$ to be an element in G with the sole restriction that $|\varphi(a_i \otimes a_j)| \leq \min(|a_i|, |a_j|)$ for all $i,j \in I$. Therefore, the ring multiplications on G are obtained by defining $a_i \cdot a_j$ to be any element in G of order $\leq \min(|a_i|, |a_j|)$ for all $i,j \in I$. The multiplication is associative (commutative) if and only if the product of basis elements of B is associative (commutative).

A straightforward approach to the above examples may be found in [36, Chapter 17]. An attempt has been made here to show that 1.2.1 may be used as an effective tool for determining the ring structures which may be defined on a group G.

1.4.7. $G = H \oplus Z(p^\infty)$, p a prime.

1. H a torsion group.

Let R be a ring with $R^+ = G$, $a \in Z(p^\infty)$, and $x \in R$. There exists $b \in Z(p^\infty)$ such that $a = |x|b$. Therefore $ax = (|x|b)x = b(|x|x) = 0$. Similarly $xa = 0$. Therefore $Z(p^\infty)$ annihilates R. For every prime $q \neq p$, H_q is an ideal in R, and $\bigoplus_{\substack{q \text{ a prime} \\ q \neq p}} H_q$ is a ring direct summand of R. It is clear from 1.4.6 how products of elements in H_q must be defined for every prime $q \neq p$.

2. H torsion free.

1) H p-divisible.

Let R be a ring with $R^+ = G$, $x \in R$, $a \in Z(p^\infty)$, with $|a| = p^k$. Clearly G is p-divisible, and so there exists $y \in G$ such that $x = p^k y$. Hence $xa = (p^k y)a = y(p^k a) = 0$. Similarly $ax = 0$, i.e., $Z(p^\infty)$ annihilates R.

If $H^2 \subseteq H$, then R is a ring direct sum $R = S \oplus T$, with $S^+ = H$, $T^+ = Z(p^\infty)$, and $T^2 = 0$. Otherwise, let $h, h_0 \in H$ such that $h \cdot h_0 = h_0' + a_0$, $h_0' \in H$, $a_0 \in Z(p^\infty)$, $a_0 \neq 0$. Then $a_0 \in \langle H \rangle$. For every positive integer n let $h_n, h_n' \in H$ such that $p^n h_n = h_0$, and $p^n h_n' = h_0'$. Put $a_n = h \cdot h_n - h_n'$. Then $a_0 = p^n a_n$, i.e. $\langle H \rangle \supseteq Z(p^\infty)$. Therefore for H p-divisible, either $\langle H \rangle \cap Z(p^\infty) = 0$, or $\langle H \rangle \supseteq Z(p^\infty)$.

b) H is not p-divisible.

Choose $b \in H$ with $h_p(b) = 0$. For any positive integer n, choose $a_n \in Z(p^\infty)$ with $|a_n| = p^n$. The map $b \otimes b \to a_n$ can be extended to an epimorphism $\varphi: H \otimes H \to Z(p^n)$. Therefore a ring R with $R^+ = G$ can be constructed so that $\langle H \rangle \cap Z(p^\infty)$ is an arbitrary proper subgroup of $Z(p^\infty)$.

Since $Z(p^\infty)$ is a direct summand of every group G containing it, example 1.4.7 is important. It will be studied further in chapters 2 and 4.

1.4.8. G a subgroup of Q^+.

The subgroups of Q^+ are, up to isomorphism, precisely the rank one torsion free groups. These groups are homogeneous. Two subgroups G_1, G_2 of Q^+ are isomorphic if and only if $t(G_1) = t(G_2)$. For every type τ, there exists a subgroup G of Q^+ such that $t(G) = \tau$, [36, Theorem 85.1].

If $t(G) = \tau$ is not idempotent, then $t(G\otimes G) = \tau^2 > \tau$ and so $Hom(G\otimes G, G) = 0$, Proposition 1.3.4, i.e., the only ring R with $R^+ = G$ is the zeroring.

Suppose that $t(G) = \tau$ is idempotent. Choose $e \in G$ such that $e \neq 0$, and $h_p(e) = 0$ or ∞ for every prime p. Let $\{q_i | i \in I\}$ be the set of primes q for which $h_q(e) = \infty$. Let R be a ring with $R^+ = G$. For $x_i \in G$ there exist integers m_i, and positive integers n_i such that $n_i x_i = m_i e$, $i = 1,2$. Therefore $n_1 n_2(x_1 \cdot x_2) = m_1 m_2(e^2)$. Since G is torsion free, if $x_i \neq 0$, $i = 1,2$, then $x_1 x_2 \neq 0$ if and only if $e^2 \neq 0$. In the latter case the product $x_1 x_2 = re^2$, $r=(m_1 m_2)/(n_1 n_2)$. Suppose that $e^2 \neq 0$. Then $e^2 = re$, $r = m/n$, m,n integers, $(m,n) = 1$. Let p be a prime such that $p \notin \{q_i | i \in I\}$. Suppose that $p|n$, i.e. $n = pk$, k an integer. Then $(p,m) = 1$, and there exist integers s,t such that $sp + tm = 1$. Therefore $ptke^2 = tne^2 = tme = e - spe$. Hence $e = p(se + tke^2)$, and $0 < h_p(e) < \infty$, a contradiction. Therefore n is the product of primes belonging to $\{q_i | i \in I\}$. Put $e' = ne$. Then $e' \neq 0$, and $h_p(e') = 0$ or ∞ for every prime p. $e'^2 = mne = me'$. It may therefore be assumed that $e^2 = me$, m a positive integer. Write $m = uv$, u,v positive integers, u a product of primes belonging to $\{q_i | i \in I\}$, and $(q,v) = 1$ for every prime $q \in \{q_i | i \in I\}$. There exists $e_1 \in G$ such that $e = ue_1$, $e_1 \neq 0$, and $h_p(e_1) = 0$ or ∞ for every prime p. $e_1^2 = u^{-2}e^2 = vu^{-1}e = ve_1$. Therefore, it may be assumed that $e^2 = me$, m a positive integer with $(q,m) = 1$ for every $q \in \{q_i | i \in I\}$.

For every positive n, let $nZ(q_i^{-1} | i \in I)$ denote the subring of Q generated by $\{n/q_i | i \in I\}$. The map $re \to mr$ for $r \in Z(q_i^{-1}; i \in I)$ is an isomorphism from R onto $mZ(q_i^{-1} | i \in I)$.

Conversely for $t(G)$ idempotent, $0 \neq e \in G$, with $t_p(e) = 0$ or ∞ for every prime p, and m a positive integer not divisible by the primes for which e has infinite height, the product $e^2 = me$ induces a ring structure on G isomorphic to $mZ(q_i^{-1} | i \in I)$, with $\{q_i | i \in I\}$ the set of primes for which $h_q(e) = \infty$.

2 Nil and quasi-nil groups

§1: Nil groups:

Definition: A group G is said to be (associative) nil if the only (associative) ring R with $R^+ = G$ is the zeroring. Associative nil groups were first investigated by T. Szele [67, 68]. The equivalence of 2) and 3) in the following theorem was proved by him [67, Satz 1 and Satz 3].

Theorem 2.1.1: Let G be a group which is not torsion free. The following are equivalent:

1) G is nil.
2) G is associative nil.
3) G is a divisible torsion group.

Proof: Clearly 1) $\Rightarrow$ 2).

2) $\Rightarrow$ 3): Suppose that G is associative nil. G is either a torsion group or a mixed group. If G is a non-divisible torsion group then $G = Z(p^k) \oplus H$, p a prime $1 \leq k < \infty$; Proposition 1.1.4 and Proposition 1.1.3. The ring direct sum of the ring of integers modulo p^k and the zeroring on H is an associative non-zeroring with additive group G, a contradiction. If G is mixed and G_t is not divisible, then again G has a cyclic direct summand, and the above argument shows that G is not associative nil. It may therefore be assumed that G_t is divisible, in which case $G = G_t \oplus H$, H a torsion free group, Proposition 1.1.2. Since $Z(p^\infty) \oplus H$ is a direct summand of G, Proposition 1.1.3, it suffices to show that $Z(p^\infty) \oplus H$ is not associative nil. Let $(a_1, a_2, a_3, \ldots, a_n, \ldots;\ pa_1 = 0,\ pa_{n+1} = a_n$ for every positive integer $n)$ be a presentation for $Z(p^\infty)$. Put $p^{-n}a_1 = a_{n+1}$ for every positive integer n. This naturally induces an action of Q on $Z(p^\infty)$. Choose $0 \neq b \in H$. Every element $g \in Z(p^\infty) \oplus H$ may be written in the form $g = d + rb + c$, with $d \in Z(p^\infty)$, $r \in Q$, $c \in H$ so that b and c are independent. Let $g_1 = d_1 + r_1 b + c_1$, $g_2 = d_2 + r_2 b + c_2$ be elements in $Z(p^\infty) \oplus H$ written in the above form. Define $g_1 \cdot g_2 = r_1 r_2 d_1$. These products induce an associative non-zeroring structure on $Z(p^\infty) \oplus H$, a contradiction.

3) $\Rightarrow$ 1): Let G be a divisible torsion group, and let R be a ring with $R^+ = G$. Let $0 \neq a$, $0 \neq b \in R$. There exists $c \in G$ sucn that $b = |a|c$. Therefore $a \cdot b = a \cdot (|a|c) = (|a| \cdot a) \cdot c = 0$.

In lieu of the above theorem, the investigation of (associative) nil groups reduces to the torsion free case. Unfortunately, not much progress has been made. The following is an up-to-date account of the results concerning torsion free nil groups.

Theorem 2.1.2: [17, Theorem 1]. Let $G = \bigoplus_{i \in I} G_i$, with G_i a homogeneous torsion free group, and $t(G_i) \cdot t(g_j) \nleq t(G_k)$ for all $i,j,k \in I$. Then G is a nil group.

Proof: By 1.2.1., $\text{Mult } G \simeq \text{Hom}(G \otimes G, G) = \text{Hom}[\bigoplus_{i,j \in I} (G_i \otimes G_j), \bigoplus_{k \in I} G_k] \subseteq \bigoplus_{i,j,k \in I} \text{Hom}(G_i \otimes G_j, G_k) = 0$.

A consequence of Theorem 2.1.2 is the following result of R. Ree and R.J. Wisner [55], (see also [41, Theorem 1.1]).

Corollary 2.1.3: Let G be a completely decomposable torsion free group, $G = \bigoplus_{i \in I} G_i$, with G_i a rank one torsion free group for each $i \in I$. Then G is (associative) nil if and only if $t(G_i) \cdot t(G_j) \nleq T(G_k)$ for all $i,j,k \in I$.

Proof: Since rank one torsion free groups are homogeneous, it suffices to show that if there exist $i_0, j_0, k_0 \in I$ such that $t(G_{i_0}) \cdot t(G_{j_0}) \leq t(G_{k_0})$, then G is not associative nil. For each $i \in I$ choose $0 \neq e_i \in G_i$. The products $e_i \cdot e_j = \begin{cases} e_{k_0} & \text{if } \{i,j\} = \{i_0, j_0\} \\ 0 & \text{otherwise} \end{cases}$ induce an associate ring structure R on G with $R^2 \neq 0$.

As a special case of Corollary 2.1.3 we have that a rank one torsion free group G is (associative) nil if and only if $t(G)$ is not indempotent, [36, Theorem 121.1]. This case however was completely investigated in Example 1.4.8.

In the cases considered so far, associativity has not influenced the classification of nil groups. So far no group G has been constructed for which the only associative ring structure which may be defined on G is the zeroring, but there exists a non-associative ring R with $R^+ = G$, and $R^2 \neq 0$. One is therefore naturally led to the following:

Conjecture 2.1.4: A Group G is nil if and only if G is associative nil.

Recently, A.E. Stratton [60, Theorem 3.3], and independently D.R. Jackett [46] made considerable progress towards classifying the nil rank two torsion free groups. Their work relies on the following result of H. Freedman [33, Theorem] which although stated here differently than in her paper [33], was in fact proved by her.

Lemma 2.1.5: Let G be a rank two torsion free group. If $T(G)$ possesses a unique minimal element, and G is not nil, then $|T(G)| \leq 3$.

Proof: For G a rank n torsion free group, the length of every chain in $T(G)$ is less than or equal n, [3, 9.1]. Let τ_0 be the unique minimal element in $T(G)$. For every $\tau \in T(G)$, either $\tau = \tau_0$ or $\tau > \tau_0$. In the latter case τ is a maximal element in $T(G)$. It therefore suffices to show that $T(G)$ possesses at most two maximal elements. Let τ_i be a maximal element in $T(G)$ and let $0 \neq x_i \in G$ with $t(x_i) = \tau_i$, $i = 1,2$, and $\tau_1 \neq \tau_2$. Let G_i be the pure subgroup of G generated by x_i, $i = 1,2$. Then G_i is a rank one torsion free group with $t(G_i) = \tau_i$, $i = 1,2$, and $G_1 \otimes G_2$ is a rank one torsion free group with $t(G_1 \otimes G_2) = \tau_1 \cdot \tau_2 > \tau_i$, $i = 1,2$, Proposition 1.3.3. The sequence:

$$0 \to G_1 \to G \to G/G_1 \to 0$$

is exact. Every torsion free group is flat, and so the sequence:

$$0 \to G_1 \otimes G_2 \to G \otimes G_2 \to (G/G_1) \otimes G_2 \to 0$$

is exact. This implies the exactness of the sequence:

$$0 \to \mathrm{Hom}[(G/G_1) \otimes G_2, G] \to \mathrm{Hom}(G \otimes G_2, G) \to \mathrm{Hom}(G_1 \otimes G_2, G)$$

[36, Theorem 44.4]. For every $0 \neq x \in G$, $t(G_1 \otimes G_2) \nleq t(x)$, and so $\mathrm{Hom}(G_1 \otimes G_2, G) = 0$ by Property 1.3.1.4. Hence $\mathrm{Hom}[(G/G_1) \otimes G_2, G] \simeq \mathrm{Hom}(G \otimes G_2, G) \simeq \mathrm{Hom}(G_2, \mathrm{End}(G))$. Let τ_3 be a maximal element in $T(G)$, $\tau_3 \neq \tau_i$, $i = 1,2$. Let $0 \neq x_3 \in G$, with $t(x_3) = \tau_3$, and put $\bar{x}$ = the coset $x + G_1$ for every $x \in G$. Clearly $t(\bar{x}_3) = t(\bar{x}_2)$, and $t(\bar{x}_3) \geq t(x_3)$. If $t(\bar{x}_3) = t(x_3)$, then

$\tau_3 = t(x_3) = t(\bar{x}_3) = t(\bar{x}_2) \geq t(x_2) = \tau_2$. By the maximality of τ_2, $\tau_3 = \tau_2$, a contradiction. Hence $\mathrm{Hom}[(G/G_1) \otimes G_2, G] = 0$ by Property 1.3.1.4) and so (A) $\mathrm{Hom}(G_2, \mathrm{End}(G)) = 0$. A similar argument shows that (B) $\mathrm{Hom}(G_1, \mathrm{End}(G)) = 0$. Let $x \in G$, and let $\varphi \in \mathrm{Hom}(G, \mathrm{End}(G))$. There exists a positive integer n, and integers n_1, n_2 such that $nx = n_1x_1 + n_2x_2$. Hence $n\varphi(x) = n_1\varphi(x_1) + n_2\varphi(x_2) = 0$ by (A) and (B). However $\mathrm{End}(G)$ is torsion free, and so $\varphi(x) = 0$, i.e., $\mathrm{Hom}(G, \mathrm{End}(G)) = 0$. By 1.2.1, G is nil, a contradiction.

The next result will show that stating the existence of a unique minimal element in the previous lemma is superfluous.

Lemma 2.1.6: Let G be a rank two torsion free group. If G is not nil then $T(G)$ possesses a unique minimal element.

Proof: Let $R = (G,\cdot)$ be a ring satisfying $R^2 \neq 0$. Suppose $x,y \in R$ with $t(x) \neq t(y)$, and $x\cdot y \neq 0$. Clearly either $t(xy) > t(x)$, or $t(xy) > t(y)$. Suppose that $t(xy) > t(x)$. Then every element in G is a linear combination of x and xy over the rationals and so $t(x)$ is the unique minimal element in G by Property 1.3.1.2). It may therefore be assumed that (A) $x\cdot y = 0$ for all $x,y \in R$ with $t(x) \neq t(y)$.

Suppose that $|T(G)| \geq 3$. Let $x \in G$. There exist $y,z \in G$ such that $t(x)$, $t(y)$, $t(z)$ are distinct types. $x = ay + bz$, with a,b rational numbers. Hence $x^2 = ax\cdot y + bx\cdot z = 0$ by (A). Let $0 \neq x$, $0 \neq y \in R$, with $t(x) = t(y)$. If x and y are independent, then every element in G is a linear combination of x, and y over the rationals, and so $t(x)$ is the unique minimal element in $T(G)$ by Property 1.3.1.2). Therefore it may be assumed that there exist non-zero integers n,m such that $nx = my$. This yields that $nmx\cdot y = n^2x^2 = 0$. Since G is torsion free $x\cdot y = 0$, or (B) $x\cdot y = 0$ for all $x,y \in R$ with $t(x) = t(y)$. Clearly (A) and (B) yield that $R^2 = 0$, a contradiction. Therefore (A) implies that $|T(G)| \leq 2$. Let $T(G) = \{\tau_1, \tau_2\}$, and choose $0 \neq x_i \in G$, $t(x_i) = \tau_i$, $i = 1,2$. Put $x = x_1 + x_2$. If $t(x) = \tau_1$ then the fact that $x_2 = x - x_1$ yields that $\tau_2 = t(x_2) \geq \tau_1$ by Property 1.3.1.2). Similarly $t(x) = \tau_2$ implies that $\tau_1 \geq \tau_2$. In either case $T(G)$ possesses a unique minimal element.

An immediate consequence of the last two lemmas is the following:

Theorem 2.1.7: Let G be a rank two torsion free group. If $|T(G)| > 3$ then G is nil.

For every positive integer n, there exist rank two torsion free groups G with $T(G) = n$, [9], and there exist rank two torsion free groups with infinite type set [36, vol. 2, p. 112, Ex. 11]. Therefore Theorem 2.1.7 is not a statement concerning the empty set.

Clearly the classification of nil rank two torsion free groups reduces to the case $|T(G)| \leq 3$. It is easy to show that the following are necessary conditions for a rank two torsion free group to be a non-nil group [60, p. 204]:

1) $|T(G)| = 1$, i.e., G homogeneous, $t(G)$ must be idempotent.
2) $|T(G)| = 2$, $T(G)$ must consist of one minimal type and one maximal type.
3) $|T(G)| = 3$, $T(G)$ must consist of one minimal type, and two maximal types.

One of the maximal types must be idempotent.

A classification of the nil rank two torsion free groups with $|T(G)| = 3$ will be given. First some preliminary results are required.

Definition: [10], and [60]. Let G be a torsion free group. The nucleus of G is

$$n(G) = \{a \in Q \mid ax \in G \text{ for all } x \in G\}.$$

Lemma 2.1.8: [60, Lemma 2.2]. Let G be a torsion free group with $r(G) > 1$. If $End(G) = n(G)$ then G is nil.

Proof: Suppose that $End(G) = n(G)$, and let R be a ring with $R^+ = G$. Suppose there exist $x,y \in R$ with $x \cdot y \neq 0$. Left multiplication by x is an endomorphism of G, hence there exists $0 \neq a \in n(G)$ such that $x \cdot z = az$ for all $z \in G$, and so $x \cdot z \neq 0$ for all $0 \neq z \in G$. Choose $z \in G$ such that x and z are independent. Right multiplication by z is also an endomorphism of G, and so $x \cdot z = bx$, $0 \neq b \in n(G)$. Hence $bx = az$ contradicting the fact that x and z are independent.

The above lemma is actually a consequence of [36, Proposition 121.2] and is in fact contained in the statement preceding [36, Proposition 121.2], namely that rigid groups of rank ≥ 2 are nil.

Lemma 2.1.9: Let G and H be torsion free groups with $G \doteq H$. Then G is (associative) nil if and only if H is (associative) nil.

Proof: It may be assumed, Proposition 1.1.6, that G and H are subgroups of a group G, and that $G \doteq H$. Suppose that H is not (associative) nil. Let $R = (H,\cdot)$ be an (associative) ring with $R^2 \neq 0$. There exists a positive integer n such that $nG \subseteq H$, and $nH \subseteq G$. For $g_1, g_2 \in G$ define $g_1 * g_2 = (ng_1)\cdot(ng_2)$. It is readily seen that $S = (G,*)$ is an (associative) ring. Let $h_1, h_2 \in H$ such that $h_1 \cdot h_2 \neq 0$. Then $(nh_1)*(nh_2) = n^4 h_1 h_2 \neq 0$. Hence $S^2 \neq 0$, and G is not (associative) nil. Similarly if G is not (associative) nil, then neither is H.

Theorem 2.1.10: Let G be a rank two torsion free group, with $|T(G)| = 3$. Then G is not (associative) nil if and only if $G \doteq G_1 \oplus G_2$, with G_i a rank one torsion free group, $i = 1,2$, and there exist $i,j,k \in \{1,2\}$ such that $t(G_i)\cdot t(G_j) \leq t(G_k)$.

Proof: Suppose that G is not strongly indecomposable, i.e., $G \doteq G_1 \oplus G_2$, with G_i a rank one torsion free group, $i = 1, 2$. By Lemma 2.1.9, G is (associative) nil if and only if $G_1 \oplus G_2$ is (associative) nil. However $G_1 \oplus G_2$ is associative nil if and only if $t(G_i)\cdot t(G_j) \nleq t(G_k)$ for all $i,j,k \in \{1,2\}$, Corollary 2.1.3.

Suppose that G is strongly indecomposable. Then $\dot{E}nd(G) = Q$, [9, Theorem 8.4]. Hence $End(G) = \{a \in Q \mid ax \in G$ for all $x \in G\} = n(G)$. Now G is nil by Lemma 2.1.8.

Note: Theorem 2.1.7 is actually implicit in [9]. As part of a complete description of the quasi-endomorphism rings of rank two torsion free groups, it was shown, [9, Theorem 8.4], that if G is a rank two torsion free group with $|T(G)| > 2$, then either $\dot{E}nd(G) = Q$, or G is quasi-isomorphic to the direct sum of two rank one groups of incomparable types. If $\dot{E}nd(G) = Q$, then G is a rigid group, and so G is nil [36, comment preceding Proposition 121.2]. If G is quasi-isomorphic to the direct sum of rank one groups of incomparable types, then it is readily seen, [9, remark preceding Theorem 8.5], that $|T(G)| = 3$.

§2. Quasi-nil Groups:

The group Q^+ admits precisely two non-isomorphic ring structures, the zeroring, and Q. Szele [68, Satz 1] classified the groups allowing precisely two non-isomorphic ring structures. He conjectured that every non-nil torsion free group $G \not\cong Q$ is the additive group of infinitely many non-isomorphic associative rings. This was shown not to be true by L. Fuchs, [38]. Borho, [12], showed that Szele's conjecture was not far from being true, see Theorem 2.2.4 and Corollary 2.2.5.

The investigation of nil groups, and additive groups of precisely two non-isomorphic rings, initiated by Szele, was generalized by Fuchs as follows:

Definition: A group G is (associative) quasi-nil if there are only finitely many non-isomorphic (associative) rings R, with $R^+ = G$.

Fuchs classified the torsion (associative) quasi-nil groups completely, and made considerable progress towards classifying the torsion free and mixed (associative) quasi-nil groups, [35], [38]. A complete description of the torsion free (associative) quasi-nil groups which are not nil was obtained by W. Borho, [11]. This in conjunction with the results of Fuchs yields a classification of the (associative) quasi-nil groups which are not nil.

The main results of this section are due to Fuchs, and Borho, [12], [35], [38].

The following technical lemma is generally useful in studying the additive groups of rings:

Lemma 2.2.1: Let R be a ring with $R^+ = G$. Then G^1 annihilates G_t.

Proof: Let $0 \neq x \in G_t$, $y \in G^1$, and let $|x| = n$. Since $y \in nG$, $y = nz$, $z \in G$. Therefore $xy = x(nz) = (nx)z = 0$. Similarly $yx = 0$.

Theorem 2.2.2: Let G be a torsion group. The following are equivalent:

1) G is quasi-nil.
2) G is associative quasi-nil.
3) $G = B \oplus D$, B a finite group, D a divisible torsion group.

Proof: Clearly 1) $\Rightarrow$ 2):

2) $\Rightarrow$ 3): Suppose that G is associative quasi-nil. $G = H \oplus D$, H reduced, and D divisible. H_p is a direct summand of G for every prime p, Proposition 1.1.1, i.e., $G = H_p \oplus K_p$. For every prime p with $H_p \neq 0$, let S_p be an associative non-zeroring, with $S_p^+ = H_p$, Theorem 2.1.1, and let T_p be the zeroring on K_p. The ring direct sum $R_p = S_p \oplus T_p$ satisfies $R_p^+ = G$, $R_p^2 \neq 0$, and $R_p \not\simeq R_q$ for distinct primes p,q. Hence $H_p = 0$ for all but finitely many primes. It therefore suffices to show that H_p is finite for every prime p. Let B_p be a basic subgroup of H_p, with basis $\{a_i | i \in I\}$. Put $|a_i| = p^{n_i}$, $i \in I$. Every set of associative products $a_i \cdot a_j$ with $|a_i \cdot a_j| \leq \min(|a_i|, |a_j|)$, $i,j \in I$, induces an associative ring structure S on H_p, [36, Theorem 120.1] and remarks following the theorem. Suppose that I is an infinite set. For every positive integer n, choose distinct elements $i_k \in I$, $k = 1,\ldots,n$.

For $i,j \in I$, define $a_i \cdot a_j = \begin{cases} 0 & \text{if } i \neq j \text{ or } i = j = i_k,\ k = 1,\ldots,n \\ a_i & \text{if } i = j, \text{ and } i \notin \{i_1, i_2,\ldots,i_n\} \end{cases}$

These products induce an associative ring structure U_n on H_p with annihilator of rank precisely n. Hence $U_n \not\simeq U_m$ for distinct positive integers n,m. Again the ring direct sums $U_n \oplus T_p$, T_p the zeroring on K_p, yield infinitely many non-isomorphic, associative ring structures on G. Therefore B_p is finite, which implies that $H_p = B_p \oplus E_p$, with E_p a divisible group. Since H_p is reduced, $H_p = B_p$, and so H_p is finite.

3) $\Rightarrow$ 1): Let $G = B \oplus D$, B a finite group, D a divisible torsion group. By Lemma 2.2.1 every ring R with $R^+ = G$ satisfies $RD = DR = 0$, and for every prime p, $B_p^2 \subseteq B_p \oplus D_p$. It may therefore be assumed that G is a p-group. Now Mult $G \simeq \mathrm{Hom}(G \otimes G, G) \simeq \mathrm{Hom}(B \otimes B, G)$, [36, Theorem 61.1]. Let $r(B) = m$. Then $r(B \otimes B) = m^2$. Let $|B| = p^n$. Every $\varphi \in \mathrm{Hom}(B \otimes B, G)$ is a homomorphic map $\varphi\colon B \otimes B \to B \oplus E[p^n]$, E a subgroup of D, with $r(E) \leq m^2$. A change of the group E is determined by an automorphism of D, and induces an isomorphic ring structure on G. Clearly $\mathrm{Hom}(B \otimes B, B \oplus E[p^n])$ is finite, and so G is quasi-nil.

Lemma 2.2.3. Let G be an (associative) torsion free quasi-nil group, and let R be an associative ring with $R^+ = G$. Then $(R^2)^+$ is a divisible group.

Proof: Put $R = R_1$, and for every positive integer n let R_n be the associative ring with $R^+ = G$, and multiplication defined by $a \times_n b = n(a \cdot b)$ for all $a,b \in G$; the product on the right hand side of the equality being the multiplication in R. Since G is (associative) quasi-nil there exist finitely many positive integers $m_1, m_2, \ldots, m_k$ such that for every positive integer n, there exists $1 \le i \le k$ such that $R_n \simeq R_{m_i}$. Put $m = \prod_{i=1}^{k} m_i$. Clearly $(m \cdot R^2)^+ = (R_m^2)^+$ is divisible, and so $(R^2)^+$ is divisible.

An alternative proof of Lemma 2.2.3 may be found in [12, Lemma 1].

Theorem 2.2.4: 1) A torsion free group is associative quasi-nil if and only if either G is nil, $G \simeq Q$, or $G \simeq Q \oplus H$, H a nil rank one torsion free group. 2) A torsion free group G is quasi-nil if and only if G is nil, or $G = Q^+$.

Proof: Let G be a torsion free associative quasi-nil group. If G is not nil, then $G = Q^+ \oplus H$, Lemma 2.2.3. If H is not reduced, then $G = Q^+ \oplus Q^+ \oplus K$. The ring direct sums of a quadratic number field on $Q^+ \oplus Q^+$ and the zeroring on K yield infinitely many non-isomorphic associative rings on G. Hence H is reduced. Suppose that $r(H) > 1$. Choose three independent elements $b_i \in G$, $i = 0,1,2$, with $b_0 \in Q$, $b_1, b_2 \in H$. Put $\tilde{G} = G \otimes Q$. Now $\tilde{G}$ may be viewed as a vector space over Q, with a natural embedding of G into $\tilde{G}$ such that b_0, b_1, b_2 are contained in a subspace $\tilde{G}_1 = b_0Q \oplus b_1Q \oplus b_2Q$ of $\tilde{G}$. Let $A = (\alpha_{ij}) \in M_2(Q)$, $i,j = 1,2$ be an arbitrary 2×2 matrix over Q. The products

$$b_i \cdot b_j = \begin{cases} \alpha_{ij} b_0 & \text{for } i,j \in \{1,2\} \\ 0 & \text{for } i = 0 \text{ or } j = 0 \end{cases}$$

induce an associative Q-algebra structure $(\tilde{G}_1)_A$ on $\tilde{G}_1$. A Q-algebra structure A_A on $\tilde{G}$ may be obtained by defining all products of an element in $\tilde{G}$ with an element in a complement of $\tilde{G}_1$ to be zero. The fact that $b_0Q \subseteq G$ yields that the above products induce an associative ring structure R_A with $R_A^+ = G$. Hence every matrix $A \in M_2(Q)$ determines an associative ring R_A, with $R_A^+ = G$. Let $0 \neq A = (\alpha_{ij})$, $0 \neq B = (\beta_{ij}) \in M_2(Q)$ with $R_A \simeq R_B$, and let $\varphi: R_A \to R_B$ be a ring isomorphism. φ is also an automorphism of G. Since H is reduced, $\varphi(b_0) \in Q$, i.e., $\varphi(b_0) = r_0 b_0$,

$0 \neq r_0 \in Q$. Now φ extends naturally to an algebra isomorphism $\tilde{\varphi}: A_A \to A_B$. Since $\tilde{\varphi}$ restricted to $(\tilde{G}_1)_A$ is a regular linear transformation

$$\tilde{\varphi}: (\tilde{G}_1)_A \to (\tilde{G}_1)_B, \ \tilde{\varphi}(b_i) = \sum_{j=0}^{2} r_{ij}b_j, \ r_{ij} \in Q, \quad i = 1,2; \quad j = 1,2.$$

$$r_0\alpha_{ij}b_0 = \tilde{\varphi}(b_i \cdot b_j) = \tilde{\varphi}(b_i)\tilde{\varphi}(b_j) = (\sum_{k=1}^{2} \sum_{\ell=1}^{2} r_{ik}r_{j\ell}\beta_{k\ell})b_0 \text{ for } i,j = 1,2.$$

Put $P = (r_{ij}) \in M_2(Q)$, $i,j = 1,2$. The last identity can be written as $r_0A = PBP^t$. Therefore the determinant $|r_0A| = r_0^2|A| = |P|^2|B|$, and so $|A| = s^2|B|$, $s \in Q$. It suffices to show that the equivalence relation in Q, $x \sim y$ if and only if $x = s^2y$, $s \in Q$, determines infinitely many equivalence classes. Suppose that the above relation partitions Q into finitely many equivalence classes. For $x \in Q$ let $\bar{x}$ be the equivalence class of x. There exist finitely many rational numbers $r_1,\ldots,r_k$ such that for every rational number r, there exists $1 \leq i \leq k$ such that $\bar{r} = \bar{r}_i$. For every integer n, the set of square roots of elements in $\bar{n}$ is precisely the field $Q(\sqrt{n})$. Therefore there are at most k quadratic number fields, a contradiction. Therefore $r(H) \leq 1$. If H is not nil let S be the zeroring on Q^+, and let T be a non-zeroring on H. Put $R = S \oplus T$. Then $(R^2)^+ = (T^2) \subseteq H$ is not divisible, contradicting Lemma 2.2.3.

Conversely, if G is nil or if $G \simeq Q^+$, then G is clearly quasi-nil. Suppose that $G = Q^+ \oplus H$, H a nil rank one torsion free group. Let $R = (G,\cdot)$ be a ring, and let π_H be the natural projection of G onto H. Define $a \times b = \pi_H(a\cdot b)$ for all $a,b \in H$. This composition defines a ring structure on H. Since H is nil, $\pi_H(a\cdot b) = 0$ for all $a,b \in H$, i.e., $H^2 \subseteq Q^+$. This coupled with the fact that, Q^+, the maximal divisible subgroup of G, is an ideal in R, yields that $R^2 \subseteq Q^+$. Choose $0 \neq b_1 \in Q$, $0 \neq b_2 \in H$. The multiplication in R is determined completely by the products $b_i \cdot b_j = \alpha_{ij}b_1$, $\alpha_{ij} \in Q$, $i,j = 1,2$. For every matrix $A = (\alpha_{ij}) \in M_2(Q)$ the above products induce a ring structure R_A on G. As above, if two such rings R_A and R_B are isomorphic, then $|A| = s^2|B|$, $s \in Q$, and so there are infinitely many isomorphic classes of ring structures on G. However if the ring R_A is required to be associative, then the equalities $(b_ib_j)b_1 = b_i(b_jb_1)$, and $(b_1b_i)b_j = b_1(b_ib_j)$ yields that

(A) $\alpha_{ij}\alpha_{11} = \alpha_{i1}\alpha_{j1}$ and (B) $\alpha_{1i}\alpha_{1j} = \alpha_{ij}\alpha_{11}$ for $i,j = 1,2$. Consider two cases a) $\alpha_{11} = 0$, b) $\alpha_{11} \neq 0$.

a) Suppose $\alpha_{11} = 0$. Putting $i = j$ in equalities (A), (B) yields that $\alpha_{i1}^2 = \alpha_{1i}^2 = 0$, and so $\alpha_{i1} = \alpha_{1i} = 0$ for $i = 1,2$. Therefore $A = \begin{pmatrix} 0 & 0 \\ 0 & \alpha \end{pmatrix}$, $\alpha \in Q$. Conversely every such matrix determines an associative ring R_A. Again map G into a vector space $\tilde{G} = b_1Q \oplus b_2Q$. The products in R_A induce a Q-algebra structure $\tilde{G}_A$ on $\tilde{G}$, with R_A a subring of $\tilde{G}_A$. The maps $b_1 \to \alpha^{-1}b_1$, $b_2 \to b_2$ induce an algebra isomorphism $\varphi: \tilde{G}_A \to \tilde{G}_{\begin{pmatrix} 0 & 0 \\ 0 & 1 \end{pmatrix}}$. Since $\tilde{G}_A^2 \subseteq R_A$, the restriction of φ to R_A is an isomorphism $\varphi: R_A \to R_{\begin{pmatrix} 0 & 0 \\ 0 & 1 \end{pmatrix}}$. Therefore every associative ring R with $R^+ = G$ is isomorphic to $R_{\begin{pmatrix} 0 & 0 \\ 0 & 0 \end{pmatrix}}$ or $R_{\begin{pmatrix} 0 & 0 \\ 0 & 1 \end{pmatrix}}$.

b) Suppose that $\alpha_{11} \neq 0$. Without loss of generality it may be assumed that $\alpha_{11} = 1$, because b_1 may be replaced by $\alpha_{11}^{-1}b_1$. Choosing $i = 1$, $j = 2$ in (A) yields that $\alpha_{12} = \alpha_{21} = \alpha$, and so choosing $i = j = 2$ in (A) yields that $\alpha_{22} = \alpha^2$. Hence if R_A is associative, then $A = \begin{pmatrix} 1 & \alpha \\ \alpha & \alpha^2 \end{pmatrix}$. The maps $b_1 \to b_1$, $b_2 \to b_2 - \alpha b_1$ induce a group automorphism of G. Now $b_1^2 = b_1$, $(b_2 - \alpha b_1)^2 = 0$, and $b_1(b_2 - \alpha b_1) = (b_2 - \alpha b_1)b_1 = 0$. Hence $\varphi: R_A \to R_{\begin{pmatrix} 1 & 0 \\ 0 & 0 \end{pmatrix}}$ is a ring isomorphism.

<u>Corollary 2.2.5</u>: Let G be a torsion free group.

1) G is the additive group of precisely two non-isomorphic (associative) rings if and only if $G \simeq Q^+$.

2) G is the additive group of precisely three non-isomorphic associative rings if and only if $G \simeq Q^+ \oplus H$, H a nil rank one torsion free group. In this case every associative ring R with $R^+ = G$ is isomorphic to one of the rings $R_{\begin{pmatrix} 0 & 0 \\ 0 & 0 \end{pmatrix}}$, $R_{\begin{pmatrix} 1 & 0 \\ 0 & 0 \end{pmatrix}}$, $R_{\begin{pmatrix} 0 & 0 \\ 0 & 1 \end{pmatrix}}$ constructed in the proof of Theorem 2.2.4, and G is the additive group of infinitely many,

non-isomorphic, non-associative rings R.

3) Otherwise G is either nil, or the additive group of infinitely many non-isomorphic associative rings.

The proof of Theorem 2.2.4 remains valid considering only commutative rings R with $R^+ = G$ by choosing symmetric matrices A in the construction of the rings R_A. Therefore:

Corollary 2.2.6: A non-nil torsion free group G is the additive group of only finitely many non-isomorphic commutative, associative rings if and only if $G \simeq Q^+$, or $G \simeq Q^+ \oplus H$, H a nil rank one torsion free group.

Theorem 2.2.7: Let G be a mixed group. G is (associative) quasi-nil if and only if either:

1) $G = B \oplus H$, B a finite group, H a torsion free (associative) quasi-nil group.

2) $G = B \oplus D \oplus H$, B a finite group, D a divisible torsion group with $D_p = 0$ for all but finitely many primes p, $H = Q^+$, or H a nil rank one torsion free group, satisfying $pH = H$ for every prime p with $D_p \neq 0$.

Proof: Suppose that G is a mixed (associative) quasi-nil group. Decompose G_t into $B \oplus D$, B reduced, and D divisible. G has a direct summand $Z(p^k)$, $0 < k < \infty$, for every prime p for which $B_p \neq 0$, Proposition 1.1.1, Proposition 1.1.4, and Proposition 1.1.3. The direct sum of the ring of integers modulo p^k on $Z(p^k)$ and the zeroring on a complement of $Z(p^k)$ yields an associative ring R_p with $R_p^+ = G$. For primes $p \neq q$, $R_p \not\simeq R_q$. Hence $B_p = 0$ for all but finitely many primes p. The same argument used in proving the implication 2) $\Rightarrow$ 3) in Theorem 2.2.2 shows that B is finite. Therefore $G = B \oplus D \oplus H$, H a torsion free group, Proposition 1.1.2. Clearly H must be (associative) quasi-nil.

Suppose that $D \neq 0$. Let p be a prime for which $D_p \neq 0$. If $pH \neq H$ then for every positive integer n, p^nH is a proper subgroup of H. Therefore there exists an epimorphism $\varphi: H \to Z(p^n)$ (the composition of the canonical homomorphism $H \to H/p^nH$ and a projection of H/p^nH onto $Z(p^n)$). Let $d \in D$ with $|d| = p^n$, and view φ as an epimorphism of H onto the subgroup of D generated by d. Let $x_i \in G$, $x_i = b_i + d_i + h_i$, $b_i \in B$,

$d_i \in D$, $h_i \in H$, $i = 1,2$. Define $x_1 \cdot x_2 = k_1 k_2 d$, with $k_i d = \varphi(h_i)$, $i = 1,2$. These products induce an associative ring structure R_n on G, in fact $R_n^3 = 0$. Now $(R_n^2)^+ \simeq Z(p^n)$, and so $R_n \not\simeq R_m$ for positive integers $n \neq m$. Hence there are infinitely many non-isomorphic associative rings with additive group G, a contradiction. Therefore $pH = H$ for all primes p for which $D_p \neq 0$.

To show that $r(H) = 1$ some facts about the endomorphisms of $Z(p^\infty)$ are required. Let $Z(p^\infty) = (a_1, a_2, \ldots, a_n, \ldots;\ pa_1 = 0,\ pa_{n+1} = a_n$ for every positive integer $n)$ be a direct summand of D. Let $\pi = \sum_{i=0}^{\infty} s_i p^i$ be a p-adic integer. For every positive integer n, define $\pi(a_n) = \sum_{i=0}^{n-1} (s_i p^i) a_n$. This action of π on a set of generators of $Z(p^\infty)$ extends naturally to an endomorphism $\pi \in \mathrm{End}(Z(p^\infty))$. In fact all the endomorphisms of $Z(p^\infty)$ are of this type, [36, vol. 1, p. 181, Example 3]. The automorphisms of $Z(p^\infty)$ are clearly p-adic units, [36, vol. 2, p. 250, Example 3].

Suppose that $r(H) > 1$. Choose independent elements $0 \neq u$, $0 \neq v \in H$. Every element $x \in G$ is of the form $x = b + d + ru + sv + w$, $b \in B$, $d \in D$, $r, s \in Q$, and $w \in H$. Let $x_i = b_i + d_i + r_i u + s_i v + w_i$, $i = 1,2$ be elements of G written in the above form. The products $x_1 \cdot x_2 = (r_1 r_2 + s_1 s_2 \pi) a_1$ induce an associative ring structure R_π on G. Put $p^\omega G = \bigcap_{n=1}^{\infty} p^n G$, and let $x \in p^\omega G$. Write x in the above canonical form, $x = b + d + ru + sv + w$. For every positive integer n, $(p^{-n}x)^2 = p^{-2n}(r^2 + s^2\pi)a_1 = (r^2 + s^2\pi)a_{2n+1}$. Let $\pi(p^\omega G)$ be the set of p-adic integers $\{r^2 + s^2\pi\}$ obtained above for each $x \in p^\omega G$, and every positive integer n. Then $\pi(p^\omega G)$ is a countable set of endomorphisms of $Z(p^\infty)$ containing the identity automorphism. If $R_\pi \simeq R_\rho$ for two p-adic integers π, ρ then the elements of $\rho(p^\omega G)$ are the elements of $\pi(p^\omega G)$ multiplied by a fixed automorphism of $Z(p^\infty)$, i.e., by a p-adic unit . However the identity automorphism must belong to $(p^\omega G)$, and so α must be chosen from a countable set of p-adic units. Therefore the isomorphism class of R_π is countable. The non-denumerability of the p-adic integers yields that there are infinitely many non-isomorphic associative rings R_π with $R_\pi^+ = G$, a contradiction. Hence $r(H) = 1$.

Let p be a prime for which $D_p \neq 0$, and let $0 \neq d \in D_p$. Let $0 \neq h \in H$. For $x_1, x_2 \in G$, put $x_i = b_i + d_i + r_i h$, $b_i \in B$, $d_i \in D$, $r_i \in Q$, $i = 1,2$. The products $x_1 x_2 = r_1 r_2 d$ induce an associative ring structure R_p on G. For primes $p \neq q$, $R_p \not\simeq R_q$. Therefore $D_p = 0$ for all but finitely many primes p.

Conversely, suppose that G is of form 1). Mult $G \simeq \mathrm{Hom}(G \otimes G, G) \simeq \mathrm{Hom}(G \otimes G, B) \oplus \mathrm{Hom}(H \otimes H, H)$, since $\mathrm{Hom}(X,Y) = 0$ for X a torsion group, and Y a torsion free group. If G is quasi-nil then so is H. Therefore $\mathrm{Hom}(H \otimes H, H) \simeq$ Mult H determines only finitely many non-isomorphic ring structures on G.
$\mathrm{Hom}(G \otimes G, B) \simeq \mathrm{Hom}(B \otimes B, B) \oplus \mathrm{Hom}(H \otimes B, B) \oplus \mathrm{Hom}(B \otimes H, B) \oplus \mathrm{Hom}(H \otimes H, B)$. The groups B, $B \otimes B$, $B \otimes H$ are finite, and so the first three summands above are finite. Since $r(H \otimes H) = 1$, Proposition 1.3.3, and B is finite, $\mathrm{Hom}(H \otimes H, B)$ is finite.

For the associative case consider $G = B \oplus H$, B finite, H an associative quasi-nil torsion free group. Let $|B| = m$. The ring multiplications on G are determined by products of elements in B, for which there are clearly only finitely many possibilities, products of elements in B with elements of H, and products of elements in H. Since mH annihilates B in every ring on G, the products of elements in B with elements of H are determined by the products of elements of B with coset representatives of elements of H/mH. Since $r(H) \leq 2$, Theorem 2.2.4.1), H/mH is finite, and there are only finitely many ways of defining products of elements of B with elements of H. Let $R = (G, \cdot)$ be an associative ring, and let $a_1, a_2 \in H$, $a_1 \cdot a_2 = a_3 + b$, $a_3 \in H$, $b \in B$. If a_i is replaced by an arbitrary element in the coset $a_i + mH$, $i = 1,2$, then the element ab above remains the same. Since there are only finitely many such cosets, it suffices to show that G/B is associative quasi-nil. However $G/B \simeq H$ is associative quasi-nil.

Suppose that G is of the form 2) with H nil. As in the case of form 1) it suffices to investigate $\mathrm{Hom}(G \otimes G, G)$. Arguments similar to those used in case 1) reduce the investigation to $\mathrm{Hom}(H \otimes H, D) \simeq \bigoplus_{p \text{ prime}} \mathrm{Hom}(H \otimes H, D_p)$ Since $D_p = 0$ for all but finitely many primes p, it suffices to consider $\mathrm{Hom}(H \otimes H, D_p)$, p a fixed prime, with $D_p \neq 0$. Now $r(H \otimes H) = 1$, and so every $\varphi \in \text{Hom}(H \otimes H, D_p)$ is a map $\varphi: H \otimes H \to Z(p^\infty)$. Changing the summand $Z(p^\infty)$ only alters the ring determined by φ by a ring isomorphism. Therefore it suffices to examine $\text{Hom}(H \otimes H, Z(p^\infty))$. However $\text{Hom}(H \otimes H, Z(p^\infty))$ is isomorphic to the additive group of p-adic numbers. If $\mu_1(x,y)$ and $\mu_2(x,y)$ are two ring multiplications on G with μ_1 a non-zero multiplication, then $\mu_2(x,y) = \pi\mu_1(x,y)$, π a p-adic number, for all $x_1, x_2 \in G$. Multiplication by p is an automorphism of $Z(p^\infty)$. Therefore, the multiplications $\mu_1(x,y)$ and $p^{-2}\rho\mu_1(x,y)$, ρ a p-adic unit, define isomorphic ring structures on G. Every non-zero p-adic number is of the form $p^k\pi$, k an integer, π a p-adic unit. Therefore the elements of $\text{Hom}(H \otimes H, Z(p^\infty))$ induce three non-isomorphic ring multiplications on G, the zero multiplication, $\mu_1(x,y)$, and $\rho\mu_1(x,y)$.

If H is not nil, then $H = Q$, and it suffices to examine the multiplications determined by $\text{Hom}(Q \otimes Q, D \oplus Q) \simeq \text{Hom}(Q \otimes Q, D) \oplus \text{Hom}(Q \otimes Q, Q)$. The second summand is isomorphic to Mult Q, and therefore determines two non-isomorphic multiplications on G. The same argument employed above to show that $\text{Hom}(H \otimes H, D)$, H nil, determines only finitely many non-isomorphic ring multiplications on G, shows the same for $\text{Hom}(Q \otimes Q, D)$. This proves Theorem 2.2.7.

3 Additive groups of nilpotent and generalized nilpotent rings

§1. The nilstufe of a group.

Another generalization of nil groups, also due to Szele is the following:

Definition: Let $G \neq 0$ be a group. The greatest positive integer n such that there exists an associative ring R with $R^+ = G$, and $R^n \neq 0$, is called the nilstufe of G, denoted $\nu(G)$. If no such positive integer n exists, then $\nu(G) = \infty$. The strong nilstufe of G, $N(G)$ is defined as above, with the associativity of R deleted.

Clearly the (associative) nil groups are precisely the groups G satisfying ($\nu(G) = 1$) $N(G) = 1$.

The following examples are due to Szele [68, Satz 2]:

Example 3.1.1: For every positive integer n, there exists a torsion free group G satisfying $\nu(G) = N(G) = n$.

For every positive integer i, let G_i be a rank one torsion free group of type $[(i,i,\ldots,i,\ldots)]$. Put $G = \bigoplus_{i=1}^{n} G_i$. Clearly $R^{n+1} = 0$ for every ring R with $R^+ = G$. For each $1 \leq i \leq n$ choose $e_i \in G_i$ such that $h(e_i) = (i,i,\ldots,i,\ldots)$. The products

$$e_i \cdot e_j = \begin{cases} e_{i+j} & \text{for } i+j \leq n \\ 0 & \text{for } i+j > n \end{cases}$$

$1 \leq i, j \leq n$ induce an associative ring structure R on G. Now $e_1^n = e_n \neq 0$, and so $R^n \neq 0$, i.e., $\nu(G) = n$. Clearly $\nu(G) \leq N(G) < n+1$, and so $N(G) = n$.

Example 3.1.2: For every positive integer n, there exists a mixed group G satisfying $\nu(G) = N(G) = n+1$.

For every positive integer i, let G_i be a rank one torsion free group of type $[(\infty,i,i,\ldots,i,\ldots)]$. Put $G = Z(2^\infty) \bigoplus_{i=1}^{n} G_i$. Clearly $R^{n+1} \subseteq Z(2^\infty)$ for every ring R with $R^+ = G$. However $Z(2^\infty)$ annihilates R, and so $R^{n+2} = 0$, i.e., $N(G) \leq n+1$. Choose $0 \neq a \in Z(2^\infty)$, and $e_i \in G_i$ with $h(e_i) = (\infty,i,i,\ldots,i,\ldots)$. The products

$$e_i \cdot e_j = \begin{cases} e_{i+j} & \text{for } i+j \leq n \\ a & \text{for } i+j = n \\ 0 & \text{for } i+j > n \end{cases}$$

for $1 \leq i, j \leq n$ induce an associative ring structure R on G. Now $e_1^{n+1} = a \neq 0$. Hence $n+1 = \nu(G) \leq N(G) \leq n+1$, and so $\nu(G) = N(G) = n+1$.

Conjecture 2.1.4 can be restated as follows: For every group G, $\nu(G) = 1$ if and only if $N(G) = 1$. It is not true in general, that $\nu(G) = N(G)$ for every group G. Examples will be given following Corollary 3.1.6.

Theorem 3.1.3: Let G be a torsion free group with $r(G) = n$, and let R be a nil associative ring with $R^+ = G$. Then $R^{n+1} = 0$.

Proof: Let $0 \neq x \in R$, and let m be the smallest positive integer such that $x^m = 0$. Suppose that $m > n+1$. Then there exist integers $a_1,\ldots,a_{m-1}$, not all zero, such that $\sum_{i=1}^{m-1} a_i x^i = 0$, i.e.,

$a_{m-1}x^{m-1} = -\sum_{i=1}^{m-2} a_i x^i$. If $a_i = 0$, $i = 1,\ldots,m-2$, then $a_{m-1} \neq 0$, and $a_{m-1}x^{m-1} = 0$. Since G is torsion free $x^{m-1} = 0$, a contradiction. Therefore $a_i \neq 0$ for some $1 \leq i \leq m-2$. Now $0 = a_{m-1}x^m = -\sum_{i=1}^{m-2} a_i x^{i+1}$, i.e., $a_{m-2}x^{m-1} = \sum_{i=1}^{m-3} a_i x^{i+1}$. Repeating the above procedure $m-3$ more time yields that $a_1 x^{m-1} = 0$, with $a_1 \neq 0$. Hence $x^{m-1} = 0$, a contradiction. Therefore $x^{n+1} = 0$ for all $x \in R$. By the Nagata-Higman Theorem

[48, p. 274], R is nilpotent. Let d be the smallest positive integer satisfying $R^d = 0$. For every positive integer k, let $(0:R^k) = \{x \in G \mid xy = 0 \text{ for all } y \in R^k\}$. Clearly $(0:R^k)$ is a pure subgroup of G for every positive integer k, and the chain $0 < (0:R^1) < (0:R^2) < \ldots < (0:R^{d-1}) = G$ is properly ascending. Hence $d \leq n+1$, [36, vol. 1, p. 120, Ex. 8 (b)].

Corollary 3.1.4: Let G be a torsion free group with $r(G) = n < \infty$. The following are equivalent:

1) Every associative ring R with $R^+ = G$ is nilpotent.

2) Every associative ring R with $R^+ = G$ is nil.

3) $\nu(G) \leq n$.

Proof: The implications 1) $\Rightarrow$ 2) and 3) $\Rightarrow$ 1) are obvious. The implication 2) $\Rightarrow$ 3) is an immediate consequence of Theorem 3.1.3.

Theorem 3.1.5: Let G be a torsion free group with $r(G) = n < \infty$. Let R be a nilpotent ring $R^+ = G$. Then $R^{2^{n-1}+1} = 0$.

Proof: Put $R^{(1)} = R$, and for every positive integer k inductively define $R^{(k+1)}$ to be the subring of R generated by elements of the form $\{ax, yb \mid a, b \in R,\ x,y \in R^{(k)}\}$. Put $V_k = R^{(k)+} \otimes Q^+$. Now V_k is a vector space over Q for every positive integer k, and $\dim_Q V_1 = n$. The descending chain $V_1 \supset V_2 \supset \ldots$ terminates at 0, and has length at most n. Hence $V_{n+1} = 0$, and so $R^{(n+1)} = 0$.

Claim: For every positive integer k, $R^{2^k+1} \subseteq R^{(k+2)}$. For $k = 0$ the claim is clearly true. Suppose $R^{2^k+1} \subseteq R^{(k+2)}$ for some positive integer k. Let $x \in R^{2^{k+1}+1}$. Then $x = y \cdot z$ with $y \in R^{2^k+1}$ or $z \in R^{2^k+1}$. Hence $y \in R^{(k+2)}$ or $z \in R^{(k+2)}$, and so $x = y \cdot z \in R^{(k+3)}$. Therefore $R^{2^{n-1}+1} \subseteq R^{(n+1)} = 0$.

Corollary 3.1.6: Let G be a torsion free group with $r(G) = n < \infty$. The following are equivalent:

1) Every ring R with $R^+ = G$ is nilpotent.

2) $N(G) \leq 2^{n-1}$.

Webb, [72, Theorem], has shown that if G is a torsion free group with $r(G) = n < \infty$, then $\nu(G) \leq n$ or $\nu(G) = \infty$, and $N(G) \leq 2^{n-1}$ or $N(G) = \infty$. These results are consequences of Corollaries 3.1.4 and 3.1.6 respectively.

Example 3.1.7: A group G satisfying $\nu(G) < N(G) < \infty$.

For $i = 1,2,3$ let G_i be a rank one torsion free group with $t(G_1) = [(1,1,\ldots,1,\ldots)]$, $t(G_2) = [(2,2,\ldots,2,\ldots)]$, and $t(G_3) = [(4,4,\ldots,4,\ldots)]$. Put $G = \bigoplus_{i=1}^{3} G_i$. Clearly $R^5 = 0$ for every ring R with $R^+ = G$. Let $e_i \in G_i$, $i = 1,2,3$, with $h(e_1) = (1,1,\ldots,1,\ldots)$, $h(e_2) = (2,2,\ldots,2,\ldots)$, and $h(e_3) = (4,4,\ldots,4,\ldots)$. The products

$$e_i \cdot e_j = \begin{cases} e_{i+j} & \text{for } i+j \leq 3 \\ 0 & \text{otherwise} \end{cases}$$

induce an associative ring structure R on G with $e_1^3 = e_3 \neq 0$. Hence $R^3 \neq 0$, and $\nu(G) = 3$ by Corollary 3.1.4.

Let e_i, $i = 1,2,3$, be as above and define

$$e_i \cdot e_j = \begin{cases} e_{2i} & \text{for } i = j < 3 \\ 0 & \text{otherwise} \end{cases}$$

These products induce a non-associative ring structure S on G with $(e_1 \cdot e_1) \cdot (e_1 \cdot e_1) = e_4$. Hence $S^4 \neq 0$, and $N(G) = 4$ by Corollary 3.1.6.

In this example the bounds of Webb's Theorem are precisely attained. For another such example see [72].

Example 3.1.8: A group G satisfying $\nu(G) < N(G) = \infty$.

For $i = 1,2$, let G_i be a rank one torsion free group with $t(G_1) = [(1,0,1,0,\ldots,1,0,\ldots)]$, and $t(G_2) = [(\infty,1,\infty,1,\ldots,\infty,1,\ldots)]$. Let $e_i \in G_i$, $i = 1,2$, with $h(e_1) = (1,0,1,0,\ldots,1,0,\ldots)$, and $h(e_2) = (\infty,1,\infty,1,\ldots,\infty,1,\ldots)$. Put $G = G_1 \oplus G_2$. It is easily verified that $\nu(G) = 2$. The products

$$e_i \cdot e_j = \begin{cases} 0 & \text{for } i = j = 2 \\ e_2 & \text{otherwise} \end{cases}$$

induce a non-associative ring structure R on G. Clearly $e_2 \in R = R^1$. Suppose that $e_2 \in R^n$. Then $e_2 = e_1 \cdot e_2 \in R^{n+1}$. Hence $R^n \neq 0$ for every positive integer n, and so $N(G) = \infty$.

An important problem in abelian group theory is to determine when a mixed group G splits into a direct sum of a torsion group and a torsion free group, [36, vol. 2, p. 185]. It is interesting to note that this occurs when $\nu(G) < \infty$, [18, Theorem 3].

Theorem 3.1.9: Let G be a mixed group. If $\nu(G) < \infty$, then $G = G_t \oplus H$, with H torsion free, and G_t the maximal divisible subgroup of G.

Proof: Let D be the maximal divisible subgroup of G. Then $G = D \oplus H$, Proposition 1.1.10. If D is not torsion then Q is a direct summand of G, Proposition 1.1.3, and so $\nu(G) = \infty$, a contradiction. Suppose that H is not torsion free. Then $Z(p^k)$ is a direct summand of G, p a prime $1 \leq k < \infty$, Proposition 1.1.4, and Proposition 1.1.3, and so $\nu(G) = \infty$, a contradiction.

Corollary 3.1.10: Let G be a mixed group. $\nu(G) < \infty$ if and only if $G = D \oplus H$, with D a divisible torsion group, and H a torsion free group with $\nu(H) < \infty$.

Proof: If $\nu(G) < \infty$, then the fact that $G = D \oplus H$, D a divisible torsion group and H torsion free with $\nu(H) < \infty$ follows almost immediately from Theorem 3.1.9.

Conversely, suppose that $G = D \oplus H$, D a divisible torsion group, and H a torsion free group with $\nu(H) = n < \infty$. Let R be an associative ring with $R^+ = G$. Clearly D is an ideal in R, and $(R/D)^+ \simeq H$. Hence $(R/D)^{n+1} = 0$, or $R^{n+1} \subseteq D$. Therefore $R^{2n+2} \subseteq D^2 = 0$ by Theorem 2.1.1, i.e., $\nu(G) \leq 2n+1$.

The problem of classifying the groups with finite nilstufe therefore reduces to the torsion free case.

Question 3.1.11: In the proof of Corollary 3.1.10 it was shown that if D is a divisible torsion group, and H is a torsion free group with $\nu(H) = n < \infty$, then $\nu(D \oplus H) \leq 2n+1$. Is it true that for $n > 1$ a positive integer, and for every positive integer k satisfying $n \leq k \leq 2n+1$, there exist a divisible torsion group D, and a torsion free group H, with $\nu(H) = n$, such that $\nu(D \oplus H) = k$?

By Corollary 3.1.6 the direct sum of n rank one torsion free groups has strong nilstufe $\leq 2^{n-1}$ or ∞. This can be improved upon as follows:

Theorem 3.1.12: Let G_i be a homogeneous torsion free group of non-idempotent type, $i = 1,\dots,n$, and let $G = \bigoplus_{i=1}^{n} G_i$. Then $\nu(G) \leq 2^n - 1$.

Proof: For $n = 1$ the theorem is true, Theorem 2.1.2. Let $n > 1$, and suppose the theorem is true for the direct sum of less than n homogeneous groups of non-idempotent type. If $t(G_i) = t(G_j)$ for $1 \leq i \neq j \leq n$, then G is the direct sum of less than n homogeneous groups of non-idempotent type, and $\nu(G) < 2^{n-1} - 1$ by the induction hypothesis. After possibly relabelling the summands G_i of G, it may be assumed that $t(G_i) \not\geq t(G_n)$ for all $1 \leq i \leq n-1$. Let R be an associative ring with $R^+ = G$, let $g \in G$, and let $g_n \in G_n$. Then $g \cdot g_n = \sum_{i=1}^{n} h_i$, with $h_i \in G_i$, $i = 1,\dots,n$. Suppose $h_i \neq 0$ for some $1 \leq i \leq n-1$. Then by Property 1.3.1.3), and Consequence 1.3.2, $t(G_i) = t(h_i) \geq t(g \cdot g_n) \geq t(g_n) = t(G_n)$, a contradiction. Hence $RG_n \subseteq G_n$, and similarly $G_nR \subseteq G_n$, i.e., G_n is an ideal in R. Put $\bar{R} = R/G_n$. By the induction hypothesis $\bar{R}^{2^{n-1}} = 0$, or $R^{2^{n-1}} \subseteq G_n$. However G_n is nil, Theorem 2.1.2, and so $R^{2^n} \subseteq G_n^2 = 0$, i.e., $\nu(G) \leq 2^n - 1$.

The conditions in Theorem 3.1.12 do not impose a bound on $N(G)$ as was seen in Example 3.1.8.

Corollary 3.1.13: Let $G = \bigoplus_{i=1}^{n} G_i$, with G_i a rank one torsion free group, $i = 1,\dots,n$. The following are equivalent.

1) $t(G_i)$ is not idempotent for all $1 \leq i \leq n$.
2) $\nu(G) \leq n$.

Proof: 1) $\Rightarrow$ 2): $\nu(G) \leq 2^n - 1$ by Theorem 3.1.13, and so $\nu(G) \leq n$ by Corollary 3.1.4.

2) $\Rightarrow$ 1): If $t(G_i)$ is idempotent for some $1 \leq i \leq n$, then G_i is the additive group of a subring of Q, 1.48, and so $\nu(G) = \infty$.

§2: Nilpotence without boundedness conditions, and generalized nilpotence.

In the previous section, groups G were considered for which there exists a positive integer n such that $R^n = 0$ for every (associative) ring R with $R^+ = G$. In this section the bound on the degree of nilpotence will be deleted. Afterwards, we will go a step further by considering the generalized nilpotent rings which were introduced by Levitzki [50].

Theorem 3.2.1: (W.J. Wickless) [74]: Let G be a group. Every (associative) ring R with $R^+ = G$ is nilpotent if and only if $G = D \oplus H$ with D a divisible torsion group, and H a reduced torsion free group such that every (associative) ring S with $S^+ = H$ is nilpotent.

Proof: Suppose that every (associative) ring R with $R^+ = G$ is nilpotent. If G_t is not divisible, then G has a non-trivial cyclic direct summand, Proposition 1.1.4 and Proposition 1.1.3, and so a non-nilpotent associative ring structure may be defined on G. Therefore $G = D \oplus H$, D divisible, and H torsion free, Proposition 1.1.2. Clearly every (associative) ring S with $S^+ = H$ must be nilpotent. If H is not reduced, then Q^+ is a direct summand of G, Proposition 1.1.3 and [36, Theorem 21.2], and so a non-nilpotent (associative) ring structure can be defined on G, a contradiction.

Conversely, suppose that $G = D \oplus H$, D a divisible torsion group, H torsion free, and such that every (associative) ring S with $S^+ = H$ is nilpotent. Let R be an (associative) ring with $R^+ = G$. Clearly D is an ideal in R, and $\bar{R} = R/D$ is nilpotent. Therefore there exists a positive integer n such that $\bar{R}^n = 0$, or $R^n \subseteq D$. Hence $R^{2n} \subseteq D^2 = 0$, Theorem 2.1.1.

Theorem 3.2.1 reduces the study of groups allowing only nilpotent (associative) ring structures to the torsion free case.

An immediate consequence of Corollaries 3.1.4 and 3.1.6 is the following:

Observation 3.2.2: Let G be a torsion free group with $r(G) < \infty$. Every (associative) ring R with $R^+ = G$ is nilpotent if and only if

$(\nu(G) < \infty)$ $N(G) < \infty$.

The proof of the following theorem of W.J. Wickless [74, Corollary to Theorem 2.1] depends on results of R.A. Beaumont and R.S. Pierce [7], [8], and will be given in Chapter 5 section 4.

Theorem 3.2.3: Let G_i be a finite rank torsion free group such that every associative ring R_i with $R_i^+ = G_i$ is nilpotent, $i = 1,\ldots,k$. Then every associative ring R with $R^+ = \bigoplus_{i=1}^{k} G_i$ is nilpotent.

Corollary 3.2.4: Let G_i be a finite rank torsion free group with $\nu(G_i) < \infty$, $i = 1,\ldots,k$. Then $\nu(\bigoplus_{i=1}^{k} G_i) < \infty$.

Proof: Theorem 3.2.3 and Observation 3.2.2.

The finiteness of the number k of summands G_i in Corollary 3.2.4 is obviously necessary. The same is true for Theorem 3.2.3.

Example 3.2.5: (Wickless [74, pp. 253-254]):

Let G_i be a rank one torsion free group with $t(G_i) = (i,i,\ldots,i,\ldots)$, and let $G = \bigoplus_{i<\omega} G_i$. For every positive integer i, let $e_i \in G_i$ with $h(e_i) = (i,i,\ldots,i,\ldots)$. The products $e_i \cdot e_j = e_{i+j}$ induce an associative ring structure R, with $R^+ = G$. For every positive integer n, $e_1^n = e_n \neq 0$, and so R is not nilpotent. However every ring R_i with $R_i^+ = G_i$, $i < \omega$, is nilpotent, Corollary 2.1.3. (The ring of polynomials constructed by Wickless, [74, pp. 253-254], is isomorphic to the above ring).

Question 3.2.6: Does Corollary 3.2.4 remain true if ν is replaced by N?

Definition: An associative ring R is said to be left (right) T-nilpotent if for every sequence $\{a_i\}_{i<\omega} \subseteq R$, there exists a positive integer n such that $a_1 \cdot a_2 \ldots a_n = 0$ $(0 = a_n \cdot a_{n-1} \ldots a_2 a_1)$. This concept was initiated by Levitzki [50], and was named T-nilpotence by Bass [4].

Theorem 3.2.7 (B.J. Gardner [42, Proposition 3.1]):

Every associative ring R with $R^+ = G$ is left T-nilpotent if and only if every associative ring R with $R^+ = G$ is right T-nilpotent.

Proof: Consider the opposite ring R^{op} of every associative ring R with $R^+ = G$. Clearly R is left T-nilpotent if and only if R^{op} is right T-nilpotent.

From now on T-nilpotence is meant to be left T-nilpotence.

Theorem 3.2.8: Let G be a group. Every associative ring R with $R^+ = G$ is T-nilpotent if and only if $G = D \oplus H$ with D a divisible torsion group, and H a reduced torsion free group such that every associative ring S with $S^+ = H$ is T-nilpotent.

Proof: Suppose that every associative ring R with $R^+ = G$ is T-nilpotent. If G_t is not divisible then G has a non-trivial cyclic direct summand, and so there exists an associative ring R with $R^+ = G$ such that R is not T-nilpotent, a contradiction. Hence $G = D \oplus H$ with D a divisible torsion group, and H a torsion free group. If H is not reduced, then Q^+ is a direct summand of G and again G allows a non-T-nilpotent associative ring structure, a contradiction.

Conversely let $G = D \oplus H$, D a divisible torsion group, and H a torsion free group such that every associative ring S with $S^+ = H$ is T-nilpotent. Let R be an associative ring with $R^+ = G$, and let $\{a_i\}_{i<\omega} \subseteq R$. For $a \in R$ let $\bar{a}$ denote the ideal class of a in $\bar{R} = R/D$. Clearly $\bar{R}$ is T-nilpotent, and so there exist positive integers $n_1 < n$ such that $\bar{a}_1 \cdot \bar{a}_2 \ldots \bar{a}_{n_1} = 0$, and $\bar{a}_{n_1+1}\bar{a}_{n+2} \ldots \bar{a}_n = 0$. Therefore $a_1 \ldots a_{n_1} \in D$, and $a_{n_1+1} \ldots a_n \in D$. However, $D^2 = 0$, Theorem 2.1.1 and so $a_1 \cdot a_2 \ldots a_n = 0$.

As was the case for groups with finite nilstufe, and groups which admit only nilpotent (associative) ring structures, the classification of groups which admit only T-nilpotent associative ring structures reduces to the torsion free case.

Observation 3.2.9: Let G_i be a finite rank torsion free group such that every associative ring R_i with $R_i^+ = G_i$ is T-nilpotent, $i = 1,\ldots,k$. Then every associative ring R with $R^+ = \bigoplus_{i=1}^{k} G_i$ is nilpotent.

Proof: Since every T-nilpotent ring is nil, every ring R_i with $R_i^+ = G_i$ is nil. By Theorem 3.1.3, every ring R_i with $R_i^+ = G_i$ is nilpotent,

$i = 1,\ldots,k$. Hence Theorem 3.2.3 yields that every ring R with $R^+ = \bigoplus_{i=1}^{k} G_i$ is nilpotent.

Another generalization of nilpotence, also due to Levitzki [50], is the following:

Definition: Let R be an associative ring. For α an arbitrary ordinal define $R^{\alpha+1}$ to be the subring of R generated by $\{xa | x \in R^{\alpha},\ a \in R\}$. For β a limit ordinal, put $R^{\beta} = \bigcap_{\alpha<\beta} R^{\alpha}$. An associative ring R is said to be α-nilpotent if $R^{\alpha} = 0$, but $R^{\beta} \neq 0$ for every ordinal $\beta < \alpha$.

A proof almost identical to that used in proving Theorem 3.2.1 and 3.2.8 shows the following:

Theorem 3.2.10: Let G be a group. For every associative ring R with $R^+ = G$ there exists an ordinal α (depending on R) such that $R^{\alpha} = 0$ if and only if $G = D \oplus H$ with D a divisible torsion group, and H a reduced torsion free group such that for every associative ring S with $S^+ = H$ there exists an ordinal β (depending on S) such that $S^{\beta} = 0$.

Question 3.2.11: Let G_i be a finite rank torsion free group, such that for every associative ring R_i with $R_i^+ = G_i$ there exists an ordinal α_i (depending on R_i) such that $R_i^{\alpha_i} = 0$, $i = 1,\ldots,k$. Let R be an associative ring with $R^+ = \bigoplus_{i=1}^{k} G_i$. Does there exist an ordinal α such that $R^{\alpha} = 0$?

The concept of α-nilpotence suggests the following:

Definition: Let G be a group. The generalized nilstufe of G, denoted $g\nu(G)$, is the first ordinal α such that $R^{\alpha} = 0$ for every associative ring R with $R^+ = G$. If no such ordinal exists, put $g\nu(G) = \infty$.

Fuchs, [36, problem 94], introduced the following:

Definition: Let G be a group. The absolute annihilator of $G = \bigcap \{(0:R) | R$ is an associative ring with $R^+ = G\}$. Here $(0:R) = \{a \in R | aR = Ra = 0\}$.

Gardner, [41, p. 386], introduced the following ascending annihilator series for a group G: Put $G(1)$ = the absolute annihilator of G. For every ordinal α, put $G/G(\alpha)(1) = G(\alpha+1)/G(\alpha)$. For β a limit ordinal, define $G(\beta) = \bigcup_{\alpha<\beta} G(\alpha)$.

<u>Theorem 3.2.12</u>: Let G be a group. If $G = G(\alpha)$ for some ordinal α, then $g\nu(G) \leq \alpha+1$.

<u>Proof</u>: Let R be an associative ring with $R^+ = G$. We wish to show that R^α is annihilated by $G(\alpha) = G$. More generally, it will be shown that $G(\beta)$ annihilates R^β for every ordinal β. For $\beta = 1$, the definition of the absolute annihilator of G assures that $G(1)$ annihilates $R = R^1$. Suppose that $G(\gamma)$ annihilates R^γ for every ordinal $\gamma < \beta$.

1) Let $\beta = \gamma+1$. Then $R^\beta \cdot G(\beta) = R^\gamma \cdot R \cdot G(\gamma+1) \subseteq R^\gamma \cdot G(\gamma) = 0$.

2) Let β be a limit ordinal, and let $a \in G(\beta)$. Then $a \in G(\gamma)$ for some $\gamma < \beta$. Hence $R^\beta \cdot a \subseteq R^\gamma \cdot a = 0$.

<u>Example 3.2.13, [23, Example 2]</u>: For every positive integer i, let G_i be a rank one torsion free group with $t(G_i) = [(i,i,\ldots,i,\ldots)]$. Put $G = \bigoplus_{i<\omega} G_i$ it is readily seen (Example 3.2.5) that $g\nu(G) > n$ for every positive integer n. Let R be an associative ring with $R^+ = G$. Every element in R^ω has type $[(\infty,\infty,\ldots,\infty,\ldots)]$ in G. However no non-zero element in G is of idempotent type. Hence $R^\omega = 0$, and $g\nu(G) = \omega$.

<u>Question 3.2.14</u>: For α an arbitrary ordinal, does there exist a group G with $g\nu(G) = \alpha$?

4 Other ring properties

§1. Semisimple, prime, semiprime, simple, division ring, field, radical ring.

Definition: Let π be a ring property. A group G is said to be an (associative) π-group if there exists an (associative) π-ring R such that $R^+ = G$. If G is not (associative) nil, and every (associative) ring R satisfying $R^2 \neq 0$, and $R^+ = G$ is a π-ring, then G is said to be an (associative) strongly π-group.

The above terminology was first employed by Beaumont and Lawver, [6], in studying the associative Jacobson semisimple, and the associative strongly Jacobson semisimple groups.

Throughout the remainder of this section, the rings considered will be assumed to be associative.

Theorem 4.1.1: Let G be a group which is not torsion free. The following are equivalent:

1) G is a field group.
2) G is a division ring group.
3) G is a simple ring group.
4) G is a prime ring group.
5) $G = \bigoplus_{\alpha} Z(p)$, p a prime, α an arbitrary cardinal.

Proof: 1) $\Rightarrow$ 2): Obvious

2) $\Rightarrow$ 3): Obvious

3) $\Rightarrow$ 4): Obvious

4) $\Rightarrow$ 5): Let R be a prime ring with $R^+ = G$. For distinct primes p,q the primary components G_p and G_q of G_t are ideals in R satisfying $G_p \cdot G_q = 0$. Hence $G_t = G_p$ for some prime p. For every positive integer k, $I = pG[p^k]$ is an ideal in R satisfying $I^k = 0$. Hence $I = 0$, and G_t is an elementary p-group, i.e., $G_t = \bigoplus_{\alpha} Z(p)$, [36, Theorem 8.5]. Therefore $G = G_t \oplus H$, with H a torsion free group,

Proposition 1.1.2. Let $x \in H$, and let $I = \langle px \rangle$. Then $G_t \cdot I = 0$, and so $I = 0$. Since H is torsion free, this implies that $x = 0$ which in turn yields that $H = 0$.

5) ⇒ 1): If G has form 5), then G is the additive group of a field extension of degree α of a field of order p, and so 5) ⇒ 1).

Definition: A ring R is Jacobson semisimple (semiprimitive) if its Jacobson radical, $J(R) = 0$. In what follows semisimple means Jacobson semisimple (semiprimitive). The reader is referred to Chapters 1 and 2 of [45] for basic facts about semisimple rings.

Theorem 4.1.2: Let G be a torsion group. The following are equivalent:

1) G is a semisimple ring group.

2) G is a semiprime ring group.

3) G is an elementary group, i.e., the order of every element in G is square order free.

Proof: 1) ⇒ 2): Let R be a semisimple ring with $R^+ = G$. Every nilpotent ideal in R is contained in $J(R) = 0$. Hence R does not possess non-zero nilpotent ideals, i.e., R is semiprime.

2) ⇒ 3): Let R be a semiprime ring with $R^+ = G$. Let p be a prime, and put $I = pG[p^2]$. Then I is an ideal in R satisfying $I^2 = 0$. Therefore $I = 0$, and so the order of every element in G is square free.

3) ⇒ 1): Let G be an elementary group. For every prime p for which $G_p \neq 0$, there exists a field F_p with $F_p^+ = G_p$, Theorem 4.1.1. Put $R = \bigoplus\{F_p \mid p$ a prime such that $G_p \neq 0\}$. Clearly, R is a semisimple ring with $R^+ = G$.

The equivalence of 1) and 3) as well as a general treatment of semisimple and strongly semisimple ring groups may be found in [6].

Theorem 4.1.3: Let G be a torsion free group. The following are equivalent:

1) G is a field group.

2) G is a division ring group.

3) G is a simple ring group

4) G is divisible, i.e., $G \simeq \bigoplus_{\alpha} Q^{+}$, α an arbitrary cardinal (Proposition 1.1.3).

Proof: The implications 1) $\Rightarrow$ 2) $\Rightarrow$ 3) are obvious.

3) $\Rightarrow$ 4): Let R be a simple ring with $R^{+} = G$. Then nR is an ideal in R for every nonzero integer n. Hence $(nR)^{+} = nG = G$, and so G is divisible.

4) $\Rightarrow$ 1): Let $G = \bigoplus_{\alpha} Q^{+}$, α an arbitrary cardinal. Then G is the additive group of a field extension of degree α of the field of rational numbers.

Notation: (see [1.4.8]): Let $\{q_i | i \in I\}$ be a set of primes. $Z(q_i^{-1} | i \in I)$ is the subring of Q generated by $\{q^{-1} | i \in I\}$.

The following theorem is due to Beaumont-Zuckerman [11], and Redei-Szele [54].

Theorem 4.1.4: Let R be a non-zeroring with R^{+} a rank one torsion free group. Then $R \simeq mZ(q_i^{-1} | i \in I)$, with m a positive integer satisfying $(m, q_i) = 1$ for all $i \in I$.

Proof: [36, Theorem 121.1], and 1.4.8.

The following results are due to Beaumont-Lawver [6].

Lemma 4.1.5: The maximal modular ideals in $mZ(q_i^{-1}|i \in I) \neq Q$ are precisely the ideals $pmZ(q_i^{-1}|i \in I)$, p a prime such that $p \notin \{q_i|i \in I\}$, and $(m,p) = 1$.

Proof: Let p be as in the statement of the lemma. $m \notin pmZ(q_i^{-1}|i \in I)$, and so $pmZ(q_i^{-1}|i \in I)$ is a proper ideal in $mZ(q_i^{-1}|i \in I)$. Let $\frac{mr}{s} \in mZ(q_i^{-1}|i \in I)$ be written in reduced form, and suppose that $\frac{mr}{s} \notin pmZ(q_i^{-1}|i \in I)$. Then $p \nmid r$. Hence there exist integers t,u such that $tr + up = 1$. Therefore $m = trm + upm$ belongs to the ideal generated by $pmZ(q_i^{-1}|i \in I)$, and $\frac{mr}{s}$, i.e., this ideal is the entire ring $mZ(q_i^{-1}|i \in I)$.

Since $(m,p) = 1$, there exist integers x,y such that $xm + py = 1$. Let $\frac{mr}{s} \in mZ(q_i^{-1}|i \in I)$. Then $xm \cdot \frac{mr}{s} = \frac{mr}{s}$ + an element in $pmZ(q_i^{-1}|i \in I)$, i.e., xm is a unity modulo $pmZ(q_i^{-1}|i \in I)$, and so $pmZ(q_i^{-1}|i \in I)$ is a modular ideal.

Let $I \neq 0$ be a proper ideal in $mZ(q_i^{-1}|i \in I) \neq Q$, and let mr be the smallest positive integer in I. Clearly $I \subseteq rmZ(q_i^{-1}|i \in I)$, and $(rm, q_i) = 1$ for all $i \in I$. Since I is a proper ideal in $mZ(q_i^{-1}|i \in I)$, $r \neq 1$. Let p be a prime such that $p|r$. Then $p \notin \{q_i|i \in I\}$, and $I \subseteq pmZ(q_i^{-1}|i \in I)$. Hence every maximal ideal M in $mZ(q_i^{-1}|i \in I)$ is of the form $M = pmZ(q_i^{-1}|i \in I)$. Suppose that M is modular. Then there exists $\frac{mr}{s} \in mZ(q_i^{-1}|i \in I)$ such that $\frac{mr}{s} \cdot m = m + \frac{pmr^1}{s^1}$ with s and s^1 products of primes in $\{q_i|i \in I\}$. Hence $mrs^1 - ss^1 = pr^1s$. Now $(ss^1, p) = 1$, and so $(m,p) = 1$. Therefore every maximal modular ideal in $mZ(q_i^{-1}|i \in I)$ is of the form $pmZ(q_i^{-1}|i \in I)$ with p a prime such that $p \notin \{q_i|i \in I\}$, and $(m,p) = 1$.

Theorem 4.1.6: Let G be a rank one torsion free group. The following are equivalent:

1) G is a strongly semisimple ring group.
2) G is a semisimple ring group.
3) $t(G) = [(k_1, k_2,\ldots,k_m,\ldots)]$ is idempotent, with $k_n = \infty$ for all n

(i.e. $G \simeq Q^+$) or $k_n = 0$ for infinitely many positive integers n.

Proof: 1) ⇒ 2): Obvious.

2) ⇒ 3): Corollary 2.1.3 and Theorem 4.1.4.

3) ⇒ 1): If $k_n = \infty$ for all positive integers n, then the one and only non-zeroring with additive group G is isomorphic to Q, and so G is a strongly semisimple ring group. Suppose that $k_n = 0$ for infinitely many positive integers n. By Corollary 2.1.3, G is not nil. Let R be a ring satisfying $R^+ = G$, and $R^2 \neq 0$. Then $R \simeq mZ(q_i^{-1} | i \in I)$, Theorem 4.1.4. The primes $\{q_i | i \in I\}$ are precisely those primes p for which $t_p(G) = \infty$. Therefore there are infinitely many primes p such that $p \notin \{q_i | i \in I\}$, and $(m,p) = 1$. Let $\frac{mr}{s} \in J(R)$, with s a product of primes in $\{q_i | i \in I\}$. Then $\frac{mr}{s} \in pmZ(q_i^{-1} | i \in I), p \notin \{q_i \; i \in I\}$, and $(m,p) = 1$, Lemma 4.1.5. Therefore infinitely many primes p satisfying $p \notin \{q_i | i \in I\}$, and $(m,p) = 1$, divide r. Hence $\frac{mr}{s} = 0$.

The equivalence of 1) and 2) in the above theorem is true in general for strongly indecomposable finite rank torsion free groups. The proof of this, and other results concerning semisimple, and strongly semisimple ring groups will be given in section 6 of this chapter, see Theorem 4.6.17. Much additional information about (strongly) semisimple ring groups may be found in [6].

A natural complement of the semisimple ring group is the radical ring group, i.e., the additive group of a ring R satisfying $R^2 \neq 0$, and $J(R) = R$. These groups have been studied by F. Haimo [44]. Some of his results, and others, will be presented here.

Lemma 4.1.7: Let G be a radical ring group. Then $G \oplus H$ is a radical ring group for every group H.

Proof: Let S be a radical ring satisfying $S^2 \neq 0$, and $S^+ = G$. Let T be the zeroring with $T^+ = H$. The ring direct sum $R = S \oplus T$, satisfies $R^2 \neq 0$, $R^+ = G \oplus H$, and $J(R) = R$.

Corollary 4.1.8: If G possesses a subgroup isomorphic to $Q^+ \oplus Q^+$, then G is a radical ring group.

Proof: $G \simeq Q^+ \oplus Q^+ \oplus H$, Proposition 1.1.10. Therefore by Lemma 4.1.7, it suffices to show that $Q^+ \oplus Q^+$ is a radical ring group. Let a,b be independent elements in $Q^+ \oplus Q^+$. The products $a^2 = b$, $ab = ba = b^2 = 0$ induce a ring structure R on $Q^+ \oplus Q^+$ satisfying $R^2 \neq 0$, but $R^3 = 0$. Hence $J(R) = R$, and $Q^+ \oplus Q^+$ is a radical ring group.

Lemma 4.1.9: If $\text{Hom}(G \otimes G, H) \neq 0$, then $G \oplus H$ is a radical ring group.

Proof: Let $0 \neq \phi \in \text{Hom}(G \otimes G, H)$. The products $(g_1+h_1) \cdot (g_2+h_2) = \phi(g_1 \otimes g_2)$ for all $g_i \in G$, $h_i \in H$, $i = 1,2$ induce a ring structure R on G satisfying $R^2 \neq 0$, but $R^3 = 0$. Hence R is associative, and $J(R) = R$, i.e., R is a radical ring.

Theorem 4.1.10: $G \oplus G$ is a radical ring group if and only if G is not nil.

Proof: If G is not nil, then $\text{Hom}(G \otimes G, G) \neq 0$, 1.2.1, and so $G \oplus G$ is a radical ring group by Lemma 4.1.9.

Suppose that G is nil. Then $\text{Mult}(G \oplus G) \simeq \text{Hom}[(G \oplus G) \otimes (G \oplus G), G \oplus G] \simeq \bigoplus_{8 \text{ copies}} \text{Hom}(G \otimes G, G) = 0$, and so $G \oplus G$ is nil.

Theorem 4.1.11: Let G be a divisible group. G is a radical ring group if and only if $G \not\simeq Q^+$ and G is not a torsion group.

Proof: A divisible torsion group is nil, Theorem 2.1.1, and Q^+ is a strongly semisimple ring group, Theorem 4.1.6. By Proposition 1.1.3 and Corollary 4.1.8 it suffices to show that $G = Q^+ \oplus D$, with D a divisible torsion group, is a radical ring group. Lemma 4.1.7 and Proposition 1.1.3 reduce the problem to showing that $Q^+ \oplus Z(p^\infty)$, p an arbitrary prime, is a radical ring group. Now $Z(p^\infty)$ is a homomorphic image of Q^+, and $Q^+ \otimes Q^+ \simeq Q^+$. Hence $\text{Hom}(Q^+ \otimes Q^+, Z(p^\infty)) \neq 0$, and $Q^+ \oplus Z(p^\infty)$ is a radical ring group by Lemma 4.1.9.

Theorem 4.1.12: Let G be a rank one torsion free group. Then G is a radical ring group if and only if t(G) is idempotent, with finitely many zero components, but not all of its components = ∞.

Proof: If t(G) does not satisfy the conditions of the theorem, then either G is nil, or G is a strongly semisimple ring group, Corollary 2.1.3, and Theorem 4.1.6.

Let G satisfy the conditions of the theorem. There is a ring R such

$R^+ = G$, and $R \simeq Z(q_i^{-1} | i \in I)$, with $\{q_i | i \in I\}$ the set of primes p, satisfying $t_p(G) = \infty$. An immediate consequence of Lemma 4.1.6 is that $J(R) = p_1 \ldots p_k R$, with $\{p_1, \ldots, p_k\}$ the set of primes p for which $t_p(G) = 0$. Now $p_1 \ldots p_k R^+ \simeq G$, so $J(R)$ is a radical ring with $[J(R)]^+ \simeq G$.

The above proof of Theorem 4.1.12 was essentially taken from the proof of [6, Theorem 3.2]. For an alternative proof, see [44, Theorem 4].

Lemma 4.1.13: Let G be a torsion group. G is a radical ring group if and only if G_p is a radical ring group for some prime p.

Proof: If G_p is a radical ring group, then so is G by Proposition 1.1.1 and Lemma 4.1.7.

Conversely, suppose that R is a radical ring with $R^+ = G$, and $R^2 \neq 0$. Then R is a ring direct sum, $R = \bigoplus_{p \text{ prime}} R_p$, with $R_p^+ = G_p$ for every prime p. Since $R^2 \neq 0$, there exists a prime p such that $R_p^2 \neq 0$. Now $R_p \lhd R$, and so $J(R_p) = R_p \cap J(R) = R_p \cap R = R_p$, i.e., G_p is a radical ring group.

Lemma 4.1.13 reduces the study of torsion radical ring groups to the p-primary case, p a prime.

Lemma 4.1.14: Let G be a p-primary group, p a prime. G is a radical ring group if and only if $G \not\simeq Z(p)$, and G is not divisible.

Proof: If G is divisible, then G is nil, Theorem 2.1.1, while the one and only non-zeroring with additive group $Z(p)$, is a field.

Suppose that $G \neq Z(p)$, and that G is not divisible. Then $G = Z(p^k) \oplus H$, k a positive integer, Propositions 1.1.4 and 1.1.3. Put $Z(p^k) = (a)$. If $H = 0$, then $k > 1$. The product $a \cdot a = pa$ induces a ring structure R on G satisfying $R^2 \neq 0$, but $R^{k+1} = 0$. Hence G is a radical ring group. If $H \neq 0$, choose $h_0 \in H$ with $|h_0| = p$. The products $a^2 = h_0$, $ah = ha = h_1 h_2 = 0$ for all $h, h_1, h_2 \in H$ induce a ring structure R on G satisfying $R^2 \neq 0$, $R^3 = 0$. Hence G is a radical ring group.

Several of the results in this section were generalized by J.D. Reid, [57], and will be presented in section 6 of this chapter.

§2: Principal ideal and Noetherian rings.

Lemma 4.2.1: Let $G = H \oplus K$, $H \neq 0$, $K \neq 0$, be an associative strongly principal ideal ring group. Then H and K are either both cyclic or both associative nil.

Proof: Suppose that H is not associative nil. Let S be an associative ring with $S^+ = H$, and $S^2 \neq 0$, and let T be the zeroring on K. The ring direct sum $R = S \oplus T$ is an associative ring satisfying $R^+ = G$, and $R^2 \neq 0$. Since T is an ideal in R, $T = \langle x \rangle$. Clearly $K = T^+ = (x)$. Therefore K is not associative nil. The above argument, interchanging the roles of H and K, yields that H is cyclic.

Corollary 4.2.2: Let $G = H \oplus K$, $H \neq 0$, $K \neq 0$ be an associative strongly principal ideal ring group. Then H and K are cyclic.

Proof: It suffices to negate that H and K are both associative nil. Let $R = (G,\cdot)$ be a ring satisfying $R^2 \neq 0$.

1) Suppose that $R^2 \subseteq K$. There exist $h_0 \in H$, $k_0 \in K$, such that $R = \langle h_0 + k_0 \rangle$. Let $h \in H$. Since $h \in R$, there exists an integer n, and $x \in R^2$ such that $h = n(h_0 + k_0) + x$. However $x \in K$. Hence $h = nh_0$, and H is cyclic, contradicting the fact that H is associative nil.

2) Suppose that $R^2 \nsubseteq K$. For all $g_1, g_2 \in G$, define $g_1 * g_2 = \pi_H(g_1 \cdot g_2)$, π_H the natural projection of G onto H. Then $S = (G,*)$ is a ring satisfying $S^2 \subseteq H$. The argument employed in 1) yields that K is cyclic, contradicting the fact that K is associative nil.

Theorem 4.2.3: Let $G \neq 0$ be a torsion group. The following are equivalent:

1) Either G is cyclic, or $G \simeq Z(p) \oplus Z(p)$, with p a prime.
2) G is a strongly principal ideal ring group.
3) G is an associative strongly principal ideal ring group.

Proof: 1) ⇒ 2): Non-trivial cyclic groups are clearly strongly principal ideal ring groups. Suppose that $G = (x_1) \oplus (x_2)$ with $|x_i| = p$ a prime, $i = 1,2$. Let R be a ring with $R^+ = G$, and $R^2 \neq 0$, and let I be a proper ideal in R. Then $|I| = 0$, or p, and so I is generated by a single element. We may assume that $R \neq \langle x_i \rangle$, $i = 1,2$. Hence $\langle x_i \rangle^+ = (x_i)$ for $i = 1,2$. This implies that

$$x_i\, x_j = \begin{cases} k_i x_i,\ 0 \le k_i < p & \text{if } i = j,\ i = 1,2, \text{ either } k_1 \ne 0 \text{ or } k_2 \ne 0. \\ 0 & \text{if } i \ne j,\ i,j = 1,2. \end{cases}$$

Put $I = \langle x_1 + x_2 \rangle$. Suppose that $k_1 \ne 0$. Let r,s be integers such that $rk_1 + sp = 1$. Then $rx_1(x_1 + x_2) = rk_1x_1 = (1-sp)x_1 = x_1$. Hence $x_1 \in I$, and so $(x_1 + x_2) - x_1 = x_2 \in I$. Therefore $I = R$. If $k_2 \ne 0$, then the above argument, reversing the roles of the indices $1,2$ again yields $I = R$.

2) $\Rightarrow$ 3): It suffices to show that G is not associative nil. This is the case by Theorem 2.1.1.

3) $\Rightarrow$ 1): Suppose that G is an associative strongly principal ideal ring group. If G is indecomposable, then $G = Z(p^n)$, p a prime, $1 \le n \le \infty$, Corollary 1.1.5. If $n = \infty$, then G is divisible, Proposition 1.1.3, and so G is nil, Theorem 2.1.1, a contradiction. Hence G is cyclic. Suppose that $G = H \oplus K$, $H \ne 0$, $K \ne 0$. By Lemma 4.2.1, either H and K are both cyclic, or both associative nil. If H and K are both associative nil, then they are both divisible, and so G is nil, Theorem 2.1.1, a contradiction. Therefore $G = (x_1) \oplus (x_2)$, with $|x_i| = n_i$, $i = 1,2$. If $(n_1, n_2) = 1$, then G is cyclic. Otherwise let p be a prime divisor of (n_1,n_2). Then $G = (y_1) \oplus (y_2) \oplus H$, with $|y_i| = p^{m_i}$, $i = 1,2$, and $1 \le m_1 \le m_2$. Since $(y_1) \oplus (y_2)$ is neither cyclic nor associative nil, $H = 0$ by Lemma 4.2.1.

The products $y_i \cdot y_j = p^{m_2-1} y_2$ for $i,j = 1,2$, induce an associative ring structure R on G with $R^2 \ne 0$. Therefore $R = \langle s_1y_1 + s_2y_2 \rangle$, s_1,s_2 integers. Every element $x \in R$ has the form

$x = k_xs_1y_1 + (k_xs_2 + m_xp^{m_2-1})y_2$, k_x, m_x integers. In particular

$y_1 = k_{y_1}s_1y_1$, and $y_2 = (k_{y_2}s_2 + m_{y_2}p^{m_2-1})y_2$. Hence if $m_2 > 1$,

$k_{y_1}s_1 \equiv 1 \pmod p$, and $k_{y_2}s_2 + m_{y_2}p^{m_2-1} \equiv 1 \pmod p$, which implies that

$p \nmid k_{y_1}$, and $p \nmid s_2$. However $k_{y_1}s_2 + m_{y_1}p^{m_2-1} \equiv 0 \pmod p$, so either

$p|k_{y_1}$ or $p|s_2$, a contradiction. Therefore $m_2 = 1 = m_1$.

Theorem 4.2.4: There are no mixed (associative) strongly principal ideal ring groups.

Proof: Let G be a mixed associative strongly principal ideal ring group. G is decomposable, Corollary 1.1.5, so by Lemma 4.2.1, $G = H \oplus K$, $H \neq 0$, $K \neq 0$, with H and K both cyclic, or both associative nil.

1) Suppose that H and K are both associative nil. There are no mixed associative nil groups, Theorem 2.1.1, so we may assume that H is a torsion group, and that K is torsion free. Let R be an associative ring with $R^+ = G$, and $R^2 \neq 0$. Clearly H is an ideal in R, and so $H = \langle h \rangle$. Let $|h| = m$. Then $mH = 0$. By Theorem 2.1.1, H is divisible, and therefore not bounded, a contradiction.

2) Suppose that $H = (x)$, and $K = (e)$ with $|x| = n < \infty$, and $|e| = \infty$. The products $x^2 = xe = ex = 0$, and $e^2 = ne$ induce an associative ring structure R on G satisfying $R^2 \neq 0$. Therefore there exist integers s,t such that $R = \langle sx + te \rangle$. Every $y \in R$ is of the form $y = m_y sx + (m_y + u_y n)te$, with m_y and u_y integers. In particular, $(m_e + u_e n)t = 1$. Hence $t = \pm 1$. Therefore $m_x + u_x n = 0$, so that $n | m_x$. However $x = m_x sx = 0$, a contradiction.

Theorem 4.2.5: Let G be a torsion free associative strongly principal ideal ring group. Then G is either indecomposable, or is the direct sum of two associative nil groups.

Proof: By Lemma 4.2.1 it suffices to negate that $G = (x_1) \oplus (x_2)$, $x_i \neq 0$, $i = 1,2$. Suppose this is so. The products

$$x_i \cdot x_j = \begin{cases} 3x_i & \text{for } i = j, \ i = 1,2 \\ 0 & \text{for } i \neq j, \ i,j = 1,2 \end{cases}$$

induce an associative ring structure R on G satisfying $R^2 \neq 0$. Therefore there exist nonzero integers k_1, k_2 such that $R = \langle k_1 x_1 + k_2 x_2 \rangle$. Every $x \in R$ is of the form $x = (r_x + 3s_x)k_1 x_1 + (r_x + 3t_x)k_2 x_2$, with r_x, s_x, t_x integers. $r_{x_1} + 3s_{x_1} = \pm 1$, so that $r_{x_1} \equiv \pm 1 \pmod 3$. However $r_{x_1} + 3t_{x_1} = 0$, so that $r_{x_1} \equiv 0 \pmod 3$, a contradiction.

Lemma 4.2.6: Let G and H be torsion free groups with $G \simeq H$. Then G is an (associative) strongly principal ideal ring group if and only if H is.

Proof: It may be assumed, without loss of generality, that $G \doteq H$. There exists a positive integer n such that $nG \subseteq H$, and $nH \subseteq G$. Suppose that G is an (associative) strongly principal ideal ring group. Let $R = (H,\cdot)$ be an (associative) ring such that $R^2 \neq 0$. The products $g_1 * g_2 = (ng_1)\cdot(ng_2)$ for all $g_1, g_2 \in G$, induce an (associative) ring structure $S = (G,*)$, with $S^2 \neq 0$. Let $I \lhd R$. Then $nI \lhd S$ and there exists $g \in G$ such that $nI = \langle g \rangle$, with $g = nh$, $h \in I$. Clearly $\langle h \rangle \subseteq I$. Let $x \in I$. Then $nx \in \langle g \rangle$, and so $nx = kg + g*y + z*g$, with k an integer, and $y, z \in G$. Therefore $nx = n[kh + (nh)\cdot(ny) + (nz)\cdot(nh)]$. Since H is torsion free, $x = kh + (nh)\cdot(ny) + (nz)\cdot(nh) \in \langle h \rangle$. Hence $I = \langle h \rangle$, and H is an (associative) strongly principal ideal ring group.

The same argument, reversing the roles of G and H, shows that if H is an (associative) strongly principal ideal ring group, then so is G.

Lemma 4.2.6 remains true if the word "strongly" is deleted.

Corollary 4.2.7: Let G be a torsion free strongly principal ideal ring group. Then G is strongly indecomposable.

Proof: Theorem 4.2.5, Corollary 4.2.2, and Lemma 4.2.6.

Theorem 4.2.8: Let G be a torsion group. The following are equivalent:

1) G is bounded.
2) G is an associative principal ideal ring group.
3) G is a principal ideal ring group.

Proof: 1) $\Rightarrow$ 2): Suppose that $nG = 0$, n a positive integer. Then $G = \bigoplus_{p|n} [\bigoplus_{\alpha_k} Z(p^k)]$, p a prime, with $p^k | n$, and α_k a cardinal number, Proposition 1.1.9. For each $p^k|n$, put $H_{p^k} = \bigoplus_{\alpha_k} Z(p^k)$. Then $G = \bigoplus_{p^k|n} H_{p^k}$.

There exists an associative principal ideal ring with unity R_{p^k}, with $R^+_{p^k} = H_{p^k}$ for all $p^k|n$, [36, Lemma 122.3]. The ring direct sum $R = \bigoplus_{p^k|n} R_{p^k}$ is an associative principal ideal ring satisfying $R^+ = G$, and $R^2 \neq 0$.

2) $\Rightarrow$ 3): Obvious.

3) $\Rightarrow$ 1): Let R be a principal ideal ring with $R^+ = G$. Then $R = \langle x \rangle$. Let $n = |x|$. Clearly $nG = 0$.

Corollary 4.2.9: Let G be a mixed group. If G is an (associative) principal ideal ring group, then G_t is bounded, and G/G_t is an (associative) principal ideal ring group. Conversely, if G_t is bounded, and if there exists a unital (associative) principal ideal ring with additive group G/G_t, then G is an (associative) principal ideal ring group.

Proof: Let R be an (associative) principal ideal ring with $R^+ = G$. Since G_t is an ideal in R, $G_t = \langle x \rangle$, and $nG_t = 0$, $n = |x|$. Now $G = G_t \oplus H$, $H \simeq G/G_t$, Proposition 1.1.2. Now $R = \langle a + y \rangle$, $a \in G_t$, $0 \neq y \in H$. Suppose that $R^2 \subseteq G_t$. Let $h \in H$. There exists an integer k_n such that $h = k_h y + b$, with $b \in R^2$. Since $R^2 \subseteq G_t$, $b = 0$, and $h = k_h y$. Therefore $H = (y)$. Clearly H is an associative principal ideal ring group. If $R^2 \nsubseteq G_t$, then $\overline{R} = R/G_t$ is an (associative) principal ideal ring with $\overline{R}^+ \simeq G/G_t$, and $\overline{R}^2 \neq 0$.

Conversely, suppose that G_t is bounded, and that there exists a unital (associative) principal ideal ring T with $T^+ = G/G_t$. Then $G \simeq G_t \oplus G/G_t$, Proposition 1.1.2. There exists a principal ideal ring S with unity, such that $S^+ = G_t$, [36, Lemma 122.3]. Let $R = S \oplus T$, with e, f the unities of S and T respectively. Let I be an ideal in R. Then $I = (I \cap S) \oplus (I \cap T)$. Now $I \cap S \lhd S$, and so $I \cap S = \langle x \rangle$. Similarly $I \cap T = \langle y \rangle$. Clearly $\langle x + y \rangle \subseteq I$. However $x = e(x+y) \in \langle x + y \rangle$ and $y = f(x+y) \in \langle x + y \rangle$. Therefore $I = \langle x + y \rangle$.

Noetherian rings are rings satisfying the ascending chain condition for left or right ideals. The following is a departure from convention:

Definition: A ring R will be said to be Noetherian, if every two sided ideal is finitely generated.

The following trivial lemma will prove to be useful.

Lemma 4.2.11. Let $G = H \oplus K$, $H \neq 0$, $K \neq 0$ be a strongly Noetherian ring group. Then either G is finitely generated, or H and K are both nil.

Proof: Suppose that H is not nil. Let S be a non-zeroring with $S^+ = H$, and let T be the zeroring with $T^+ = K$. The ring direct sum $R = S \oplus T$ satisfies $R^+ = G$, and $R^2 \neq 0$. Since $T \lhd R$, $T = \langle t_1, \ldots, t_n \rangle$.

Clearly $T = (t_1,\dots,t_n)$, and so T is not nil. The same argument reversing the roles of H and K yields that H is finitely generated. Hence G is finitely generated.

Theorem 4.2.12: Let G be a non-torsion free group. G is a strongly Noetherian ring group if and only if G is finitely generated.

Proof: By Lemma 4.2.10, it suffices to show that if G is a strongly Noetherian ring group, then G is finitely generated.

1) Suppose that G is a torsion group. If G is indecomposable, then $G \simeq Z(p^n)$, p a prime, $1 \leq n \leq \infty$, Corollary 1.1.5. If $n = \infty$, then G is nil, Proposition 1.1.3, and Theorem 2.1.1, a contradiction. Hence H is cyclic. If $G = H \oplus K$, $H \neq 0$, $K \neq 0$, and if G is not finitely generated, then by Lemma 4.2.11 both H and K are nil, and so G is nil by Theorem 2.1.1, a contradiction.

2) Suppose that G is mixed. Then G is not indecomposable, Corollary 1.1.5, and so $G = H \oplus K$, $H \neq 0$, $K \neq 0$. If G is not finitely generated, then by Lemma 4.2.11, both H and K are nil. By Theorem 2.1.1 it may be assumed that H is a torsion group, and that K is torsion free. Let R be a Noetherian ring with $R^+ = G$. Since $H \lhd R$, $H = \langle h_1,\dots,h_k \rangle$. Let $n = \prod_{i=1}^{k} |h_i|$. Then $nH = 0$. By Theorem 2.1.1, H is divisible, a contradiction.

Corollary 4.2.13: Let G be a strongly Noetherian ring group. Then G_t and G/G_t are strongly Noetherian ring groups.

Proof: If G is torsion free, the statement of the corollary is trivial. Otherwise G is finitely generated by Theorem 4.2.12, and so G_t and G/G_t are strongly Noetherian ring groups by Lemma 4.2.10.

A torsion free strongly Noetherian ring group need not be finitely generated; e.g., Q^+. However, we have the following:

Theorem 4.2.14: Let G be a strongly Noetherian ring group. Then G is either indecomposable, or finitely generated.

Proof: By Lemma 4.2.11, it suffices to negate that $G = H \oplus K$, $H \neq 0$, $K \neq 0$, with H and K both nil. Suppose this is so. Let R be a ring with $R^+ = G$, and $R^2 \neq 0$. Then $R = \langle x_1, \ldots, x_n \rangle$, n a positive integer. Put $x_i = h_i + k_i$, $i = 1, \ldots, n$.

1) Suppose that $R^2 \subseteq K$. Let $h \in H$. There exist integers $r_1, \ldots, r_n$, and $x \in R^2$ such that $h = \sum_{i=1}^{n} r_i(h_i + k_i) + x$. However $R^2 \subseteq K$. Hence $h = \sum_{i=1}^{n} r_i h_i$, and H is finitely generated. This contradicts the fact that H is nil.

2) Suppose that $R^2 \nsubseteq K$. Let π_H be the natural projection of G onto H. The products $g_1 * g_2 = \pi_H(g_1 \cdot g_2)$ for all $g_1, g_2 \in G$ induce a ring structure S on G such that $0 \neq S^2 \subseteq H$. The argument used in 1) shows that K is finitely generated, contradicting the fact that K is nil.

For a discussion of the additive groups of Noetherian rings in accordance with the usual definition, see [36, section 126].

Fuchs, [36, Problem 97], asked whether or not free groups of large cardinality allow a Noetherian ring structure (Noetherian meaning left Noetherian, although the problem is also open for Noetherian according to the definition given here). He noted [36, vol. 2, p. 311], that no Noetherian ring is known whose additive group is free and non-denumerable. Such a ring has been constructed by J.D. O'Neill, [52, Theorem 1], as follows:

Let ω_1 be the first non-denumerable ordinal. Put $A = Z[x_i | i \in I]$, the commutative associative ring of polynomials in indeterminates $\{x_i | i \in I\}$, I = the set of ordinals less than ω_1. Let S = the set of elements in A which are relatively prime to every positive integer. The localization A_S is a Noetherian ring. The proof of this fact may be found in [52, Theorem 1].

§3. Descending chain conditions for ideals.

The additive groups of (left) Artinian rings have been completely classified by Fuchs and Szele [70] as follows:

Proposition 4.3.1: A group G is the additive group of a (left) (associative) Artinian ring if and only if $G \simeq \bigoplus_{\alpha} Q^+ \bigoplus_{\text{finite}} Z(p_i^\infty) \bigoplus_{\beta} Z(p_j^{k_j})$ with $p_j^{k_j} | m$, where α, β are arbitrary cardinals, p_i, p_j are primes, and m is a fixed positive integer.

Proof: Let R be an Artinian ring. $R \supseteq 2!\, R \supseteq 3!\, R \supseteq \ldots$ descending chain of left ideals in R. Hence there exists a positive integer k such that $n!R = k!R$ for all $n \geq k$. Put $m = k!$ Clearly $(mR)^+$ is divisible, and so $(mR)^+ \simeq \bigoplus_{\alpha} Q \bigoplus_{p \text{ a prime}} \bigoplus_{\alpha_p} Z(p^\infty)$. Clearly, $R^+/(mR)^+$ is bounded, and so $R^+ \simeq \bigoplus_{\alpha} Q^+ \bigoplus_{p \text{ a prime}} \bigoplus_{\alpha_p} Z(p^\infty) \bigoplus \bigoplus_{\beta} Z(p_j^k)$, p_j primes, and $p_j^k | m$, m a fixed positive integer, Proposition 1.1.9. Put $D = \bigoplus_{p \text{ a prime}} \bigoplus_{\alpha_p} Z(p^\infty)$. Clearly, $D \subseteq (R^+)^1$, and so D annihilates R_t^+, Lemma 2.1. The complement of R_t^+ in R^+ is isomorphic to $\bigoplus_{\alpha} Q^+$ which is clearly annihilated by D. Hence D belongs to the annihilator of R, and so every subgroup of D is an ideal in R. If D possesses infinitely many non-trivial direct summands D_i, $i = 1,2,\ldots,$ then $\bigoplus_{i=1}^{\infty} D_i \supset \bigoplus_{i=2}^{\infty} D_i \supset \ldots \supset \bigoplus_{i=n}^{\infty} D_i \supset \ldots$ is a properly descending infinite chain of (left) ideals in R, a contradiction. Therefore $D = \bigoplus_{\text{finite}} Z(p_i^\infty)$, p_i primes.

Conversely, let $G \simeq \bigoplus_{\alpha} Q^+ \bigoplus_{\text{finite}} Z(p_i^\infty) \bigoplus Z(p_j^{k_j})$ with $p_j^{k_j}$ m, where α, β are arbitrary cardinals; p_i, p_j are primes; and m is a fixed positive integer. Let F be a field with $F^+ \simeq \bigoplus_{\alpha} Q^+$, S a zeroring with $S^+ = \bigoplus_{\text{finite}} Z(p_i^\infty)$. Now $\bigoplus_{\beta} Z(p_j^{k_j}) = \bigoplus_{j=1}^{n} \bigoplus_{k=1}^{n_j} \bigoplus_{\beta_j} Z(p_j^k)$, p_j a prime, and $p_j^k | m$; $j = 1,\ldots,n$; $k = 1,\ldots,n_j$. For β_j finite, let T_{jk} be the sum of

β_j copies of the ring of integers modulo p_j^k. For β_j infinite, an associative commutative ring with unity T_{jk} can be constructed so that $T_{jk}^+ = \bigoplus_j Z(p_j^k)$, and the only (left) ideals in T_{jk} are $T_{jk}, pT_{jk}, \ldots, p^{n_j}T_{jk} = 0$, [36, Lemma 122.3]. Put $T = \bigoplus_{j=1}^{n} \bigoplus_{k=1}^{n_j} T_{jk}$. The ring $R = F \oplus S \oplus T$ is Artinian with $R^+ = G$.

The ring R constructed above is commutative, associative, and possesses only finitely many ideals. These facts will prove to be useful later.

In this section it will be shown that the additive groups of rings satisfying the descending chain condition for two sided ideals are precisely the additive groups of Artinian rings described above. The additive groups of rings possessing only finitely many ideals will be studied. The structure of the additive group of a ring possessing only finitely many ideals (satisfying the descending chain condition for two sided ideals) will be employed to obtain necessary and sufficient conditions for embedding the ring into a ring with unity possessing only finitely many ideals (satisfying the descending chain condition for two sided ideals).

For the remainder of this section, ideal will mean two sided ideal and the "descending chain condition" will be abbreviated DCC.

<u>Theorem 4.3.2</u>. Let G be a torsion free group.

The following are equivalent:

1) G is the additive group of an (associative) ring possessing only finitely many ideals.

2) G is the additive group of an (associative) ring satisfying the DCC for ideals.

3) $G \simeq \bigoplus_{\alpha} Q^+$, α an arbitrary cardinal.

<u>Proof</u>: Clearly 1) $\Rightarrow$ 2) and 3) $\Rightarrow$ 1), the latter implication since $\bigoplus Q^+$ is the additive group of a field. It therefore suffices to show that:

2) $\Rightarrow$ 3): Let R be a ring satisfying the DCC for ideals, with $R^+ = G$. For every prime p, and positive integer n, $p^nR \supseteq p^{n+1}R$. Hence there exists a positive integer n such that $p^nR = p^{n+1}R$. Therefore for $a \in G$, there

exists $b \in G$ such that $p^n a = p^{n+1} b$, or $p^n(a - pb) = 0$. Since G is torsion free, $a = pb$, and so G is p-divisible for every prime p. Hence G is divisible, and $G \simeq \bigoplus_\alpha Q^+$, Proposition 1.1.3.

Observe that the above theorem adds two equivalent conditions to those given in Theorem 4.1.3.

Corollary 4.3.3: Let G be a non-nil torsion free group. The following are equivalent:

1) Every (associative) ring R with $R^+ = G$, and $R^2 \neq 0$ possesses only finitely many ideals.

2) Every (associative) ring R with $R^+ = G$, and $R^2 \neq 0$ satisfies the DCC for ideals.

3) $G \simeq Q^+$.

Proof: Again the implications 1) ⇒ 2) and 3) ⇒ 1) are obvious.

2) ⇒ 3): Suppose condition 2) is satisfied. By Theorem 4.3.2, $G \simeq \bigoplus_\alpha Q^+$, and so $G \simeq Q^+ \oplus H$. Let S be the zeroring with $S^+ = H$. The ring direct sum $R = Q \oplus S$ satisfies $R^+ \simeq G$, and $R^2 \neq 0$. If $H \neq 0$, choose $0 \neq a \in H$. The infinite chain of ideals in R, $(a) \supset (2!a) \supset (3!a) \supset \ldots$ is properly descending, a contradiction. Hence $H = 0$, and $G \simeq Q^+$.

Lemma 4.3.4: Let G be a torsion group. The following are equivalent:

1) G is the additive group of an (associative) ring possessing only finitely many ideals.

2) G is the additive group of an (associative) ring satisfying the ascending chain condition for ideals.

3) $G = \bigoplus_{i=1}^{m} \bigoplus_{j=1}^{n_i} \bigoplus_{\alpha_j} Z(p_i^j)$, p_i a prime, m, n_i positive integers, α_j an arbitrary cardinal, $i = 1, \ldots, m$; $j = 1, \ldots, n_i$.

Proof: Clearly 1) ⇒ 2):

2) ⇒ 3): Suppose there exist infinitely many distinct primes $\{p_i\}_{i=1}^{\infty}$ for which $G_{p_i} \neq 0$. Let R be an (associative) ring with $R^+ = G$ such that R satisfies the ascending chain condition for ideals. The infinite chain of

ideals in R, $G_{p_1} \subset \bigoplus_{i=1}^{2} G_{p_i} \subset \ldots \subset \bigoplus_{i=1}^{k} G_{p_i} \subset \ldots$ is properly ascending, a contradiction. Hence $G_p \neq 0$ for only a finite set of primes $\{p_i\}_{i=1}^{m}$. For every $p \in \{p_i\}_{i=1}^{m}$, $G_p[p] \subseteq G_p[p^2] \subseteq \ldots G_p[p^k] \subseteq \ldots$ is an ascending chain of ideals in R, each contained in G_p. Hence there exists a positive integer n_i such that $G_{p_i} = G_{p_i}[p_i^{n_i}]$, $i = 1,\ldots,m$. Therefore

$$G_{p_i} = \bigoplus_{j=1}^{n_i} \bigoplus_{\alpha_j} Z(p_i^j),\ \alpha_j \text{ a cardinal number, and so } G = \bigoplus_{i=1}^{m} \bigoplus_{j=1}^{n_i} \bigoplus_{\alpha_j} Z(p_i^j).$$

3) ⇒ 1): The ring T constructed in the proof of Proposition 4.3.1 is an associative, commutative ring possessing only finitely many ideals, and $T^+ = G$.

Clearly, G satisfies condition 2) of Lemma 4.3.4 if and only if G is a Noetherian ring group in the sense of the definition in section 2 of this chapter. Condition 3) merely states that G is bounded. Therefore by Theorem 4.2.8 we have the following:

Theorem 4.3.5: Let G be a torsion group. The following are equivalent:

1) G is bounded.

2) G is an associative principal ideal ring group.

3) G is a principal ideal ring group.

4) G is the additive group of an associative ring satisfying the ascending chain condition for ideals.

5) G is the additive group of a ring satisfying the ascending chain condition for ideals.

6) G is the additive group of an associative ring possessing only finitely many ideals.

7) G is the additive group of a ring possessing only finitely many ideals.

Corollary 4.3.6: Let G be a non-nil torsion group.

The following are equivalent:

1) Every (associative) ring R with $R^+ = G$, and $R^2 \neq 0$ possesses only finitely many ideals.

2) $G = \bigoplus_{i=1}^{m} \bigoplus_{j=1}^{n_i} \bigoplus_{\alpha_j} Z(p_i^j)$, p_i a prime, n, n_i positive integers, α_j a finite cardinal, $i = 1,\dots,m$, $j = 1,\dots,n_i$ (i.e. G is finite).

Proof: 1) ⇒ 2): Suppose G satisfies condition 1). It may be assumed that G is of the form 3) of Lemma 4.3.4. Suppose that α_j is an infinite cardinal for some index j. Then $G = \bigoplus_{i=1}^{\infty} (a_i) \oplus H$, $|a_i| = p^j$, p a prime. Let R_i be a ring isomorphic to the ring of integers modulo p^j, with $R_i^+ = (a_i)$, $i = 1,2,\dots$, and let S be the zeroring with $S^+ = H$. The ring direct sum $R = \bigoplus_{i=1}^{\infty} R_i \oplus S$ satisfies $R^+ = G$, and $R^2 \neq 0$. $R_i \triangleleft R$ for $i = 1,2,\dots$, a contradiction.

2) ⇒ 1): Obvious.

Theorem 4.3.7: Let G be a torsion group. The following are equivalent:

1) G is the additive group of an (associative) ring satisfying the DCC for ideals.

2) $G = \bigoplus_{i=1}^{m} \bigoplus_{j=1}^{n_i} \bigoplus_{\alpha_j} Z(p_i^j) \bigoplus_{\text{finite}} Z(p_i^\infty)$, p_i a prime, m, n_i positive integers, α_j an arbitrary cardinal, $i = 1,\dots,m$, $j = 1,\dots,n_i$.

Proof: 1) ⇒ 2): Let R be a ring satisfying the DCC for ideals with $R^+ = G$. Suppose there exists an infinite set of primes $\{p_i\}_{i=1}^{\infty}$ such that $G_{p_i} \neq 0$, $i = 1,2,\dots$ The infinite chain of ideals in R, $\bigoplus_{i=1}^{\infty} G_{p_i} \supset \bigoplus_{i=2}^{\infty} G_{p_i} \supset \dots \supset \bigoplus_{i=k}^{\infty} \supset \dots$ is properly descending, a contradiction. Hence $G_p \neq 0$ for only finitely many primes $\{p_i\}_{i=1}^{m}$. Let $p \in \{p_i\}_{i=1}^{m}$, and let D_p be the maximal divisible subgroup of G_p. Now $D_p = \bigoplus_{\alpha} Z(p^\infty)$. Suppose that α is an infinite cardinal. Then R possesses a properly descending chain of ideals $\bigoplus_{i=1}^{\infty} Z(p^\infty) \supset \bigoplus_{i=2}^{\infty} Z(p^\infty) \supset \dots \supset \bigoplus_{i=k}^{\infty} Z(p^\infty) \supset \dots$, a contradiction. $G_p = H_p \oplus D_p$, H_p a reduced subgroup of G_p,

Proposition 1.1.10. For every positive integer n, $p^nG_p \supseteq p^{n+1}G_p$ is an inclusion of ideals in R. Hence there exists a positive integer n, such that $p^nG_p = p^{n+1}G_p$. Clearly $p^kG_p = p^kH_p \oplus D_p$ for every positive integer k. Therefore $p^nH_p = p^{n+1}H_p$. Since H_p is reduced, this implies that $p^nH_p = 0$, i.e., for every $1 \leq i \leq m$, there exists a positive integer n_i such that $H_{p_i} = \bigoplus_{j=1}^{n_i} \bigoplus_{\alpha_j} Z(p_i^j)$, α_j a cardinal number, $j = 1,\ldots,n_i$.

Clearly G is of the form of condition 2).

2) $\Rightarrow$ 1): If G satisfies condition 2) then G is the additive group of an associative ring satisfying the DCC for left ideals, Proposition 4.3.1.

Corollary 4.3.8: Let G be a torsion group. G is the additive group of a ring with trivial annihilator satisfying the DCC for ideals if and only if G satisfies one, and hence all, of the equivalent conditions in Theorem 4.3.5.

Proof: If G is the additive group of a ring with trivial annihilator, then G is reduced, Lemma 2.2.1. If in addition G is the additive group of a ring satisfying the DCC for ideals, then G is bounded by Theorem 4.3.7.

Conversely, if G satisfies the equivalent conditions of Theorem 4.3.5, then the ring constructed in proving the implication 3) $\Rightarrow$ 1) for Lemma 4.3.4, is a ring with trivial annihilator satisfying the DCC for ideals.

Corollary 4.3.9: Let G be a non-nil torsion group. The following are equivalent:

1) Every (associative) ring R with $R^+ = G$, and $R^2 \neq 0$ satisfies the DCC for ideals.

2) $G = \bigoplus_{i=1}^{m} \bigoplus_{j=1}^{n_i} \bigoplus_{\alpha_j} Z(p_i^j) \bigoplus_{\text{finite}} Z(p_i^\infty)$, p_i a prime, n, n_i positive integers, α_j a finite cardinal, $i = 1,\ldots,m$, $j = 1,\ldots,n_i$.

Proof: Similar to the proof of Corollary 4.3.6.

Theorem 4.3.10: Let G be a group. The following are equivalent:

1) G is the additive group of an (associative) ring possessing only

finitely many ideals.

2) $G \simeq \bigoplus_{\alpha} Q^{+} \bigoplus_{i=1}^{m} \bigoplus_{j=1}^{n_i} \bigoplus_{\alpha_j} Z(p_i^j)$, p_i a prime, m, n_i non-negative integers, α, α_j arbitrary cardinals, $i = 1,\ldots,m$; $j = 1,\ldots,n_i$.

Proof: 1) ⇒ 2): Let R be a ring possessing only finitely many ideals, with $R^+ = G$. Then $\overline{R} = R/G_t$ is a ring possessing only finitely many ideals, with $\overline{R}^+$ torsion free. Therefore $\overline{R}^+ \simeq \bigoplus_{\alpha} Q^+$, α an arbitrary cardinal, Theorem 4.3.2. Let $I \lhd G_t$, $0 \neq a \in I$, $x \in R$. Since G/G_t is divisible, there exist $y \in G$, $b \in G_t$ such that $x = |a|y + b$. Clearly $xa = ba$, and $ax = ab$. Hence I is an ideal in R, and G_t is of the form of condition 3) in Lemma 4.3.4. By Proposition 1.1.2, $G = H \oplus G_t$, H torsion free. Clearly $H \simeq G/G_t \simeq \bigoplus_{\alpha} Q^+$.

2) ⇒ 1): Let G be a group satisfying condition 2). Define a field structure F on $\bigoplus_{\alpha} Q^+$, and let R be the ring constructed in proving the implication 3) ⇒ 1) for Theorem 4.3.2. The ring direct sum $S = F \oplus R$ is associative, possesses only finitely many ideals, and $S^+ \simeq G$.

Corollary 4.3.11: Let G be a non-nil group. The following are equivalent:

1) Every (associative) ring R with $R^+ = G$, and $R^2 \neq 0$ possesses only finitely many ideals.

2) $G \simeq \bigoplus_{\alpha} Q^{+} \bigoplus_{i=1}^{m} \bigoplus_{j=1}^{n_i} \bigoplus_{\alpha_j} Z(p_i^j)$, p_i a prime, m, n_i non-negative integers, α, α_j finite cardinals, $i = 1,\ldots,m$; $j = 1,\ldots,n_i$.

An argument almost identical to that used in proving Theorem 4.3.1 yields:

Theorem 4.3.12: Let G be a group. The following are equivalent:

1) G is the additive group of an (associative) ring satisfying the DCC for ideals.

2) $G \simeq \bigoplus_{\alpha} Q^{+} \bigoplus_{i=1}^{n} \bigoplus_{j=1}^{n_i} \bigoplus_{\alpha_j} Z(p_i^j) \bigoplus_{\text{finite}} Z(p_i^{\infty})$, p_i a prime, m, n_i non-negative integers, α, α_j arbitrary cardinals, $i = 1,\ldots,m$;

$j = 1,\dots,n_i$.

The following corollaries are easy to prove:

Corollary 4.3.13: Let G be a group. G is the additive group of an (associative) ring with trivial annihilator, satisfying the DCC for ideals if and only if G satisfies one, and hence both of the equivalent conditions of Theorem 4.3.10.

Corollary 4.3.14: Let G be a non-nil group. The following are equivalent:

1) Every ring R with $R^+ = G$, and $R^2 \neq 0$ satisfies the DCC for ideals.

2) $G \simeq \bigoplus_{\alpha} Q^+ \bigoplus_{i=1}^{m} \bigoplus_{j=1}^{n_i} \bigoplus_{\alpha_j} Z(p_i^j) \bigoplus_{\text{finite}} Z(p_i^\infty)$, p_i a prime, m, n_i non-negative integers, α, α_j finite cardinals, $i = 1,\dots,m$; $j = 1,\dots,n_i$.

Lemma 4.3.15: Let R be an (associative) ring with trivial annihilator, such that $R^+ = D \oplus (R^+)_t$, D a divisible group. Then R is a ring direct sum $R = S \oplus T$, with $S^+ = D$, and $T^+ = (R^+)_t$.

Proof: Let $x \in D$, $0 \neq a \in (R^+)_t$. Then $x = |a|x_1$, $x_1 \in D$, and so $ax = xa = 0$. Hence $D(R^+)_t = (R^+)_t D = 0$, and it suffices to show that $D^2 \subseteq D$. Let $x,y \in D$. Then $xy = z+a$, $z \in D$, $a \in (R^+)_t$. Suppose that $a \neq 0$. Since $a \in (R^+)_t$, a annihilates D. Let $0 \neq b \in (R^+)_t$. There exist x_1, $z_1 \in D$ such that $x = |b|x_1$, $z = |b|z_1$. Therefore $ab = [|b|x_1 y - |b|z_1]b = 0$. Similarly $ba = 0$. Hence a annihilates R, a contradiction.

Corollary 4.3.16: Let R be an (associative) ring with trivial annihilator satisfying the DCC for ideals. Then R is the direct sum of a torsion free ring and a torsion ring.

Proof: Theorem 4.3.12 and Lemma 4.3.15.

An argument almost identical to that used in proving [36, Theorem 123.1] yields the following results:

Theorem 4.3.17: An (associative) ring R possessing only finitely many ideals can be embedded in an (associative) ring with unity possessing only finitely many ideals.

Theorem 4.3.18: An (associative) ring R satisfying the DCC for ideals can be embedded in an (associative) ring with unity satisfying the DCC for ideals if and only if R^+ has no summand $Z(p^\infty)$, i.e., $(R^+)_t$ is reduced.

Corollary 4.3.19: An (associative) ring with trivial annihilator satisfying the DCC for ideals, can be embedded in an (associative) ring with unity satisfying the DCC for ideals.

Proof: Corollary 4.3.13 and Theorem 4.3.18.

Definition: A ring R satisfying the DCC for (right) principal ideals is called an (MHR)MHI ring.

The MHR and MHI rings were introduced by F.A. Szasz, [63], [64], [65], and [66]. They play an important role in the theory of radicals of rings.

Most of the results presented here, are taken from [64], and [31].

Theorem 4.3.20: Let G be a group. The following are equivalent:

1) G is an MHR ring group.
2) G is an MHI ring group.
3) $G = D \oplus K$, with D a divisible group, and K a reduced torsion group.

Proof: Clearly 1) $\Rightarrow$ 2).

2) $\Rightarrow$ 3): Let R be an MHI ring with $R^+ = G$. Let D be the maximal divisible subgroup of G. Then $G = D \oplus K$, with K a reduced group, Proposition 1.1.10. Let $a \in K$. Then $\langle a\rangle \supseteq \langle 2!a\rangle \supseteq \langle 3!a\rangle \supseteq \ldots$ is a descending chain of principal ideals in R. Hence there exists a positive integer m such that $\langle n!a\rangle = \langle m!a\rangle$ for all $n \geq m$. Clearly $\langle m!a\rangle^+$ is divisible, and so $\langle m!a\rangle^+ \subseteq D$. This obviously implies that $m!a = 0$, or that K is a torsion group.

3) $\Rightarrow$ 1): Let $G = D \oplus K$, with D a divisible group, and K a reduced torsion group. By Proposition 1.1.3, there exists a torsion free divisible group D_0 such that $G = D_0 \oplus G_t$. If $D_0 = 0$, then the zeroring with additive group G is an MHR ring. If $D_0 \neq 0$, let F be a field with $F^+ = D_0$, Theorem 4.1.3, and let S be the zeroring with $S^+ = G_t$. The ring direct sum $R = F \oplus S$ is an MHR ring with $R^+ = G$.

Corollary 4.3.21: Let R be an MHI ring, and let D be the maximal divisible subgroup of R^+. Then D_t annihilates R.

Proof: By the above theorem, $R^+ = D \oplus K$, with D a divisible group, and K a reduced torsion group. D annihilates K by Lemma 2.2.1, and so D_t annihilates K. Let $a \in D$, $b \in D_t$. Then $a = |b|c$ for some $c \in D$. Hence $ab = (|b|c)b = c(|b|b) = 0$. Similarly $ba = 0$, i.e., D_t annihilates D, which yields that D_t annihilates R.

Theorem 4.3.22: Let G be a group. The following are equivalent:

1) G is a unital MHR ring group.
2) G is a unital MHI ring group.
3) $G = D \oplus B$, with D a torsion free divisible group, and B a bounded group.

Proof: Clearly 1) $\Rightarrow$ 2).

2) $\Rightarrow$ 3): Theorem 4.3.20 and Corollary 4.3.21 imply that $G = D \oplus B$, with D a torsion free divisible group, and B a reduced torsion group. Let R be a unital MHI ring with $R^+ = G$, and let e be the unity in R. Then $e = e_1 + e_2$, $e_1 \in D$, $e_2 \in B$. For $x \in B$, $x = ex = e_1x + e_2x$. Let n be a positive integer such that $nx = 0$. Since D is divisible, $e_1 = nd$, $d \in D$. Therefore $e_1x = (nd)\cdot x = d\cdot(nx) = 0$. Therefore $x = e_2x$, and so $|e_2|x = (|e_2|e_2)x = 0$, i.e., B is bounded.

3) $\Rightarrow$ 1): Suppose that $G = D \oplus B$, with D a torsion free divisible group, and B a bounded group. Then $G = D \oplus \bigoplus_{i=1}^{m} \bigoplus_{\alpha_i} Z(n_i)$, m, n_i positive integers, and α_i an arbitrary cardinal, $i = 1,\ldots,m$, Proposition 1.1.9. Let F be a field with $F^+ = D$, Theorem 4.1.3. Then the ring direct sum $R = F \oplus \bigoplus_{i=1}^{m} [\bigoplus_{\alpha_i} Z/n_iZ]$ is a unital MHR ring with $R^+ \simeq G$.

Corollary 4.3.23: Let R be a unital MHI (MHR) ring. Then $R = R_0 \oplus R_t$, with R_0 a torsion free unital MHI (MHR) ring, and R_t a bounded unital MHI (MHR) ring.

Proof: $R^+ = D \oplus R_t$, with D a torsion free divisible group, and R_t bounded, Theorem 4.3.22. Clearly R_t is an ideal in R. Let n be a positive integer such that $nR_t = 0$. Then $D\cdot R_t = (nD)\cdot R_t = D\cdot(nR_t) = 0$.

Similarly, $R_t \cdot D = 0$. Let $a \in R$. Left (right) multiplication by a induces a homomorphism $D \to R^+$ via $x \to ax$ $(x \to xa)$. Hence aD^+ (Da^+) is divisible, and so $aD \subseteq D$ $(Da \subseteq D)$, i.e., D is an ideal in R. Put $R_0 = (D,\cdot)$. Clearly $R = R_0 \oplus R_t$ is a ring direct sum. Let e be the unity in R. Then $e = e_0 + e_1$, with e_0 the unity in R_0, and e_1 the unity in R_t. It is readily seen that R_0 and R_t are MHI (MHR) rings.

Lemma 4.3.24: Let R be an MHI ring, and let I be an ideal in R with I^+ torsion free. Then I^+ is divisible.

Proof: Let $a \in I$. Then $\langle a\rangle \supseteq \langle 2!a\rangle \supseteq \langle 3!a\rangle \supseteq \ldots$ is a descending chain of principal ideals in R. Therefore there exists a positive integer m such that $\langle k!a\rangle = \langle m!a\rangle$ for every integer $k \geq m$. Clearly $\langle m!a\rangle^+$ is divisible. Therefore for every positive integer n there exists $x \in \langle m!a\rangle$ such that $m!a = nm!x$, or $m!(a - nx)$. Since I^+ is torsion free, $a = nx$, and I^+ is divisible.

Theorem 4.3.25: An MHI (MHR) ring R can be embedded into a unital MHI (MHR) ring if and only if R_t is bounded.

Proof: Clearly if R is a subring of a unital MHI (MHR) ring, then R_t is bounded by Theorem 4.3.22. Conversely, let R be an MHI (MHR) ring with R_t bounded. By Theorem 4.3.20, $R^+ = D \oplus R_t$, with D a torsion free divisible group. Put $R_0 = (D,\cdot)$. The same argument used in proving Corollary 4.3.23 shows that $R = R_0 \oplus R_t$ is a ring direct sum of MHI (MHR) rings. Put $H = R_0^+ \oplus Q^+$. For $a_i \in R_0$, $q_i \in Q$, $i = 1,2$, define $(a_1 + q_1)\cdot(a_2 + q_2) = a_1 \cdot a_2 + q_1a_2 + q_2a_1 + q_1q_2$, where the first product, $a_1 \cdot a_2$, is defined by the multiplication in R_0, the second and third products by the natural action of Q on R_0 (e.g. if $q_1 = n/m$, then $q_1a_2 = b \in R_0$ such that $mb = na_2$), and the fourth product by multiplication in Q. Then $S_0 = (H,\cdot)$ is a unital MHI (MHR) ring with $R_0 < S_0$. Let n be a positive integer such that $nR_t = 0$. Put $K = R_t^+ \oplus (Z/nZ)^+$. Defining multiplication on K in a manner similar to that in H yields a unital MHI (MHR) ring structure T with $T^+ = K$, and $R_t < T$. Put $S = S_0 \oplus T$. Then S is a unital MHI (MHR) ring, and $R \triangleleft S$.

The technique employed above in embedding R_0 and R_t into unital MHI (MHR) rings is an application of a general embedding procedure, (see 36, Lemma 123.2).

Observation 4.3.26: Let R be an MHI (MHR) ring. If R can be embedded as a subring into a unital MHI (MHR) ring, then R can be embedded as an ideal into a unital MHI (MHR) ring.

F.A. Szasz asked for a description of the MHR rings which may be embedded into a unital MHR ring. The problem was solved in [31].

Additive groups of rings have played an important role in the solution of several embedding problems for rings; see [39], [70], and [24].

§4. Subdirectly irreducible rings.

Definition: A ring R is subdirectly irreducible if the intersection of its nonzero ideals is nonzero.

Lemma 4.4.1: Let G be an (associative) subdirectly irreducible ring group. Then G_t is p-primary for some prime p.

Proof: If $G_p = 0$ for every prime p, then $G_t = 0$. Otherwise let p be a prime for which $G_p \neq 0$. Let R be an (associative) subdirectly irreducible ring with $R^+ = G$. For every prime $q \neq p$, $G_p \cap G_q = 0$. Since G_p and G_q are ideals in R, $G_q = 0$.

Theorem 4.4.2: Let G be a torsion group. G is an (associative) subdirectly irreducible ring group if and only if G is a bounded p-primary group, or G is a p-primary group with $G' = Z(p^k)$, $1 \leq k \leq \infty$, p a prime.

Proof: Suppose that G is a subdirectly irreducible ring group. G is p-primary for some prime p by Lemma 4.4.1. Let R be a subdirectly irreducible ring with $R^+ = G$. Suppose that $G' = 0$. Then $\bigcap_{n<\omega} p^nG = 0$. Since p^nG is an ideal in R for every positive integer n, there exists a positive integer n such that $p^nG = 0$, i.e., G is bounded.

Suppose that $G' \neq 0$. For every subgroup H of G', $RH = HR = 0$, Lemma 2.2.1, and so H is an ideal in R. If $G' = H \oplus K$, then H and K are ideals in R with $H \cap K = 0$, i.e., G' is indecomposable. By Corollary 1.1.5, $G' = Z(p^k)$, $1 \leq k \leq \infty$.

Let G be a bounded p-primary group with basis $\{a_i \mid i \in I\}$, $|a_i| = p^{k_i}$, $k_i \leq k$ for all $i \in I$. Choose $i_0 \in I$ such that $k_{i_0} = k$. For $i,j \in I$

define

$$a_i \cdot a_j = \begin{cases} p^{k-k_i} a_{i_0} & \text{for } i = j \\ 0 & \text{for } i \neq j . \end{cases}$$

These products induce an associative ring structure R on G. Let $0 \neq x \in R$. Then $x = n_1 a_{i_1} + \ldots + n_r a_{i_r}$, i_j distinct elements of I, and n_j an integer with $p^{k_{i_j}} \nmid n_j$ for $j = 1,\ldots,r$. Now $x a_{i_1} = n_1 p^{k-k_{i_1}} a_{i_0}$. Therefore every nonzero ideal in R contains $p^{k-1} a_{i_0}$, and so R is subdirectly irreducible.

Let G be a p-primary group with $G' = Z(p^r)$, $1 \leq r \leq \infty$. Then G' possesses a unique minimal subgroup generated by some $a_0 \in G'$, $|a_0| = p$. For every positive integer k, choose $a_k \in G$ such that $p^k a_k = a_0$. Let B be a basic subgroup of G with basis $\{b_i | i \in I\}$, $|b_i| = p^{k_i}$ for all $i \in I$. For $i,j \in I$ define

$$b_i \cdot b_j = \begin{cases} a_{k_i - 1} & \text{for } i = j \\ 0 & \text{for } i \neq j . \end{cases}$$

These products induce an associative ring structure R on G, [36, Theorem 120.1]. Let $a \in R$, $a \neq 0$, $|a| = p^j$. Now $B = \bigoplus_{n<\omega} B_n$, with $B_n = \bigoplus_{\alpha_n} Z(p^n)$, α_n a cardinal number, and B_n a direct summand of G for all $n < \omega$. Let r be the smallest positive integer such that under the natural projection $\pi_r\colon G \to B_r$, $\pi_r(a) \neq 0$. Let n be a positive integer, $n \geq j+r$. Since

$$G = \bigoplus_{1 \leq i \leq n} B_i \quad \bigoplus_{n+1 \leq i < \omega} B_i + p^n G,$$

$a = m_1 b_{i_1} + \ldots + m_s b_{i_s} + g$, with i_t distinct elements of I, m_t an integer, $t = 1,\ldots,s$, $k_{i_1} = r$, $m_1 = p^u v$, u,v integers, $u < r$,

$(p,v) = 1$, and $g \in p^r G$. Now $a \cdot p^{r-u-1} b_{i_1} = va_0$. Therefore every nonzero ideal in R contains a_0.

Observation 4.4.3: Let G be a mixed (associative) subdirectly irreducible ring group. Then G_t is a p-primary group possessing nonzero elements of infinite height, p a prime.

Proof: G_t is p-primary by Lemma 4.4.1. Let R be an (associative) subdirectly irreducible ring, with $R^+ = G$. For every positive integer n, $p^n G_t$ is an ideal in R. If $\bigcap_{n<\omega} p^n G_t \neq 0$, then G possesses elements of infinite height. Otherwise $p^n G_t = 0$ for some positive integer n. In the latter case G_t and $p^n G$ are nonzero ideals in R satisfying $G_t \cap p^n G = 0$, a contradiction.

Theorem 4.4.4: Let G be a torsion free group. G is an (associative) subdirectly irreducible ring group if and only if G is not reduced.

Proof: Suppose that G is an (associative) subdirectly irreducible ring group. Let R be a subdirectly irreducible ring with $R^+ = G$. The αth Ulm subgroup G^α of G is an ideal in R for every ordinal α. Let α be the smallest ordinal for which $G = G^{\alpha+1}$. If α is a limit ordinal, then $G^\alpha = \bigcap_{\beta<\alpha} G^\beta$ is the intersection of nonzero ideals in R, and so $G^\alpha \neq 0$. If α is an isolated ordinal, say $\alpha = \beta+1$, then $G^\alpha = \bigcap_{n<\omega} nG^\beta$ is the intersection of nonzero ideals nG^β of R, and so $G^\alpha \neq 0$. Therefore G^α is a nonzero divisible subgroup of G.

Suppose that G is not reduced. Then $G = D \oplus H$, D a divisible group, H a reduced group, $D \neq 0$. Let $\{e_i \mid i \in I\}$ be a maximal independent set for D, and let $\{a_j \mid j \in J\}$ be a maximal independent set for H. Choose $i_0 \in I$. Define $e_i e_j = e_i a_k = a_k e_i = a_k a_\ell = e_{i_0}$ for all $i,j \in I$, $k,\ell \in J$. These products induce an associative ring structure on G. Every nonzero ideal in R contains e_{i_0}, and so R is subdirectly irreducible.

Lemma 4.4.5: Let G be an (associative) strongly subdirectly irreducible ring group, $G = H \oplus K$, $H \neq 0$, $K \neq 0$. Then H and K are both (associative) nil.

Proof: Suppose that H is not (associative) nil. Let S be an (associative) ring with $S^+ = H$, $S^2 \neq 0$, and let T be the zeroring, with $T^+ = K$. The ring direct sum $R = S \oplus T$ is an (associative) ring satisfying $R^+ = G$, and $R^2 \neq 0$. Now S and T are ideals in R, with $S \cap T = 0$, a contradiction.

Theorem 4.4.6: Let G be a torsion free group. G is an (associative) strongly subdirectly irreducible ring group if and only if $G \simeq Q^+$.

Proof: $G \simeq Q^+ \oplus H$ by Theorem 4.4.4. Hence $G \simeq Q^+$ by Lemma 4.4.5. The converse is obvious.

Theorem 4.4.7: Let G be a torsion group. G is an (associative) strongly subdirectly irreducible ring group if and only if $G = Z(p^k)$, p a prime, k a positive integer.

Proof: Suppose that G is an (associative) strongly subdirectly irreducible ring group. If $G = H \oplus K$, $H \neq 0$, $K \neq 0$, then H and K are (associative) nil, Lemma 4.4.5, and so G is nil by Theorem 2.1.1, a contradiction. Therefore $G = Z(p^k)$, p a prime, $1 \leq k \leq \infty$, Corollary 1.1.5. Since G is not nil, $k < \infty$ by Theorem 2.1.1. Clearly $Z(p^k)$, p a prime, $1 \leq k < \infty$, is a strongly subdirectly irreducible ring group.

Theorem 4.4.8: Let G be a mixed group. G is a strongly subdirectly irreducible ring group if and only if $G = Z(p^\infty) \oplus H$, H a rank one, p-divisible, torsion free, nil group.

Proof: Suppose that G is a strongly subdirectly irreducible ring group. $G = Z(p^\infty) \oplus H$, H a nil torsion free group, Proposition 1.1.4, Lemma 4.4.5, and Theorem 2.1.1. Suppose that $r(H) = \alpha > 1$. Then $r(H \otimes H) = \alpha^2 > 1$, [36, vol. 1, p. 261, Ex. 8]. Let D be the divisible hull of $H \otimes H$. Then $D = Q_1 \oplus K$, $Q_1 \simeq Q^+$, $K \neq 0$. Let $x \in Q_1$, $x \neq 0$. There exists a positive integer m such that $mx \in H \otimes H$. Now $m = p^k n$, $(p,n) = 1$. Let $a \in Z(p^\infty)$, $|a| > p^k$. There exists a homomorphism $\varphi: Q_1 \to Z(p^\infty)$ such that $\varphi(x) = a$. Now $\varphi(mx) = m\varphi(x) \neq 0$. Let π be the natural projection of D onto Q_1, and put $\psi = \varphi \cdot \pi$. Then ψ_H, the restriction of ψ to $H \otimes H$, is a nonzero homomorphism. Let $g_i = a_i + b_i$, $a_i \in Z(p^\infty)$, $b_i \in H$, $i = 1,2$. Define $g_1 * g_2 = \psi_H(b_1 \otimes b_2)$. Then $R = (G,*)$ is a ring with $R^+ = G$, and $R^2 \neq 0$ Let $h \in H$ be such that $h \otimes h \in (H \otimes H) \cap K$. Let L be the subgroup of H

generated by h. Then $Z(p^\infty)$ and L are nonzero ideals in R, with $Z(p^\infty) \cap L = 0$, a contradiction. Therefore $r(H) = 1$.

It now suffices to show that H is p-divisible. If not, there exists $h \in H$ such that $h_p(h) = 0$. For $a,b \in G$, define $a \cdot b = 0$ if $a \in Z(p^\infty)$ or $b \in Z(p^\infty)$. Choose $a_0 \in Z(p^\infty)$ with $|a_0| = p$. For $h_i \in H$ there exists a positive integer n_i, and an integer m_i such that $p \nmid n_i$, and $n_i h_i = m_i h$, $i = 1,2$. There exists a unique element $a \in Z(p^\infty)$ such that $n_1 n_2 a = m_1 m_2 a_0$. Define $h_1 \cdot h_2 = a$. The above products induce a ring structure R on G with $R^2 \neq 0$. Now $Z(p^\infty)$ and pH are ideals in R with $Z(p^\infty) \cap pH = 0$, a contradiction.

Conversely, let $G = Z(p^\infty) \oplus H$, H a rank one, p-divisible, torsion free, nil group. Clearly G is p-divisible. G is not nil by Theorem 2.1.1. Let R be a ring with $R^+ = G$, and $R^2 \neq 0$. Since H is nil, $R/Z(p^\infty)$ is a zeroring, i.e., $R^2 \subseteq Z(p^\infty)$. Let $a_0 \in Z(p^\infty)$, $|a_0| = p$. Every nonzero subgroup of $Z(p^\infty)$ contains a_0. Let I be an ideal in R, $I \neq 0$. Suppose that $a_0 \notin I$. Then $I \cap Z(p^\infty) = 0$. However $RI \subseteq R^2 \subseteq Z(p^\infty)$, and so $RI = 0$, and similarly $IR = 0$. Let $0 \neq x \in I$, $x = a+b$, $a \in Z(p^\infty)$, $b \in H$. Clearly $b \neq 0$. There exists a positive integer n such that $p^n a = 0$. Hence $p^n x \in H \cap I$. We may therefore assume that there exists $0 \neq h_0 \in H \cap I$. Let $h \in H$. There exist a non-negative integer k, a positive integer r, and an integer s such that $p \nmid r$, and $p^k rh = sh_0$. Therefore $p^k rhR \subseteq IR = 0$. However $p^k rhR = rh(p^k R) = rhR$. Now $hR \subseteq Z(p^\infty)$, and for $0 \neq a \in Z(p^\infty)$, $ra \neq 0$. Hence $hR = 0$. Therefore 1) $HR = 0$, and similarly 2) $RH = 0$.

Let $a \in Z(p^\infty)$, $x \in R$. There exists a positive integer n such that $p^n a = 0$, and there exists $y \in R$ such that $x = p^n y$. Hence $ax = a(p^n y) = (p^n a)y = 0$. Therefore 3) $Z(p^\infty) \cdot R = 0$, and similarly 4) $R \cdot Z(p^\infty) = 0$. Equalities 1), 2), 3), and 4) imply that $R^2 = 0$, a contradiction.

§5. Local rings.

In this section a ring is meant to be an associative ring.

Definition: A unital ring R is local, if its nonunits form an ideal. This ideal is clearly the unique maximal ideal in R.

Notation:

Q^* $\quad Q - \{0\}$.

Q^p $\quad \{a/b \mid a,b \in Z,\ p \nmid a,\ p \nmid b\}$.

F_p $\quad$ a field of order p, p a prime.

Lemma 4.5.1: Let R be a local ring with maximal ideal M, and set of units U.

1) If $(R/M)^+ \simeq \bigoplus_\alpha Q^+$, α an arbitrary cardinal, then $Q^* \subseteq U$.

2) If R^+ is not a torsion group, and if $(R/M)^+ = \bigoplus_\alpha Z(p)$, α an arbitrary cardinal, then $Q_p \subseteq U$.

3) If R^+ is a torsion group, then $F_p \subseteq U$.

Proof: 1) $1+M$ is torsion free in $(R/M)^+$, and so $n = n\cdot 1 \notin M$ for every nonzero integer n. Hence n, and $1/n$ belong to U, and so $\frac{n}{m} = n(\frac{1}{m}) \in U$ for arbitrary nonzero integers n,m.

2) follows from the same argument as for 1), assuming n and m to be relatively prime to p.

3) again follows from the same argument, plus the fact that $p(1+M) = M$, i.e., $p \in M$.

Observe that R/M above is a simple ring, so either $(R/M)^+ \simeq \bigoplus_\alpha Q^+$, or $(R/M)^+ = \bigoplus_\alpha Z(p)$, Theorems 4.1.1 and 4.1.3. Therefore the conditions imposed on $(R/M)^+$ in Lemma 4.5.1 and in the following lemma are not restrictive.

Lemma 4.5.2: Let R be a local ring with maximal ideal M, and $R^+ = G$.

1) If $(R/M)^+ \simeq \bigoplus_\alpha Q^+$, α an arbitrary cardinal, then G is torsion free.

2) If $(R/M)^+ = \bigoplus_\alpha Z(p)$, α an arbitrary cardinal, then G_t is p-primary.

Proof: 1) Suppose that $(R/M)^+ \simeq \bigoplus_\alpha Q^+$. It suffices to show that $G_q = 0$ for every prime q. Let $x \in G_q$, $|x| = q^k$. By Lemma 4.5.1, $q^k, q^{-k} \in R$, and so $x = q^{-k}q^k x = 0$.

2) Suppose that $(R/M)^+ = \bigoplus_\alpha Z(p)$. Let q be a prime, $q \neq p$, and let $x \in G_q$, $|x| = q^k$. By Lemma 4.5.1, $q^k, q^{-k} \in R$, and so $x = q^{-k} \cdot q^k x = 0$.

Theorem 4.5.3: Let G be a torsion group. G is a local ring group if and only if $G = \bigoplus_{k=1}^{n} \bigoplus_{\alpha_k} Z(p^k)$, p a prime, n a positive integer, α_k an arbitrary cardinal, k = 1,...,n.

Proof: Let R be a local ring with $R^+ = G$. Let $|1| = n$. Clearly $nx = 0$ for all $x \in G$. Hence G is a bounded group, and so G is a direct sum of cyclic groups, Proposition 1.1.9. G is p-primary, Lemma 4.5.2, and so $G = \bigoplus_{k=1}^{n} \bigoplus_{\alpha_k} Z(p^k)$. Conversely, let $G = \bigoplus_{k=1}^{n} \bigoplus_{\alpha_k} Z(p^k)$. Put $H = \bigoplus_{\alpha_n} Z(p^n)$. If α_n is an infinite cardinal, then there exists a local ring T, with $T^+ = H$, [36, Lemma 122.3]. If $\alpha_n = r < \infty$, then $H = (a_1) \oplus \ldots \oplus (a_r)$, $|a_i| = p^n$, $i = 1,\ldots,r$. The products $a_1 \cdot a_j = a_j \cdot a_1 = a_j$ for $j = 1,\ldots,r$, and $a_i \cdot a_j = pa_1$ for $i,j = 2,\ldots,r$, induce a ring structure T on H with unique maximal ideal $(pa_1) \oplus (a_2) \oplus \ldots \oplus (a_r)$. In either case, the unity $e \in T$ is a basis element for H, i.e., $H = (e) \bigoplus_{i \in I} (a_i)$. Let $L = \bigoplus_{k=1}^{n-1} \bigoplus_{\alpha_k} Z(p^k)$, and let $\{b_j \mid j \in J\}$ be a basis for L. Define $eb_j = b_j e = b_j$, and $b_j \cdot b_k = b_k \cdot b_j = a_i \cdot b_j = b_j \cdot a_i = 0$ for all $i \in I$, $j,k \in J$. Define the product of elements in H in accordance with the multiplication in T. These products induce a ring structure R on G. Let N be the maximal ideal in T. Then $M = N \oplus L$ is the unique maximal ideal in R.

Theorem 4.5.4: G is the additive group of a local ring R with maximal ideal M such that $(R/M)^+ \simeq \bigoplus_{\alpha} Q^+$, α an arbitrary cardinal, if and only if $G \simeq \bigoplus Q^+$.

Proof: Let R be a local ring with $(R/M)^+ \simeq \bigoplus_{\alpha} Q^+$. Let $x \in G$ and let n be a positive integer. Since n is a unit in R, Lemma 4.5.1, $x = n(\frac{1}{n}x)$, and so G is divisible. G is torsion free by Lemma 4.5.2. Therefore $G \simeq \bigoplus Q^+$, Proposition 1.1.3.

Conversely, if $G \simeq \bigoplus_{\alpha} Q^+$, then G is the additive group of a field.

Theorem 4.5.5: Let R be a local ring with maximal ideal M. If

$(R/M)^+ = \bigoplus_{\alpha} Z(p)$, α an arbitrary cardinal, and if R^+ is not a torsion group, then $R^+ = H \oplus K$, H a divisible group and K homogeneous of type $[(\infty,\ldots,1,\infty,\ldots)]$, with 1 at the p-th component.

Proof: Let q be a prime, $q \neq p$. By Lemma 4.5.1, $q, q^{-1} \in R$. Hence for every $x \in R^+$, $x = q(q^{-1}x)$, and so R^+ is q-divisible. Let H be the maximal divisible subgroup of R^+. Then $R^+ = H \oplus K$, K homogeneous of type $[(\infty,\ldots,1,\infty,\ldots)]$ with 1 at the p-th component.

Theorem 4.5.6: Let R be a Noetherian local ring. Then $R^+ = H \oplus \bigoplus_{k=1}^{n} \bigoplus_{\alpha_k} Z(p^k)$, n a positive integer, p a prime, α_k an arbitrary cardinal, $k = 1,\ldots,n$, and H torsion free. If R^+ is mixed, then $pH \neq H$.

Proof: R_t^+ is p-primary for some prime p, Lemma 4.5.2. Since $R^+[p] \subseteq R^+[p^2] \subseteq \ldots$ is an ascending chain of ideals in R, $R_t^+ = G[p^n]$ for some positive integer n, i.e., $R_t^+ = \bigoplus_{k=1}^{n} \bigoplus_{\alpha_k} Z(p^k)$. Therefore $R^+ = H \oplus \bigoplus_{k=1}^{n} \bigoplus_{\alpha_k} Z(p^k)$, with H torsion free, Proposition 1.1.2.

Let M be the maximal ideal in R. If R^+ is a mixed group, then R_t^+ and $p^nR = p^nH$ are proper ideals in R. Hence $R_t^+ \subseteq M$, and $p^nH \subseteq M$. If $p^nH = H$, then $H \oplus R_t^+ = R^+ \subseteq M$, a contradiction. Therefore $pH \neq H$.

§6: Rings with trivial left annihilator, subrings of algebraic number fields, and semisimple rings continued.

This section consists of the work of J.D. Reid, [56], and [57]. Several results in section one of this chapter are special cases of results which will now be obtained. The above approach was advantageous in that only elementary techniques were employed. The approach in this section requires somewhat more sophisticated techniques, but yields more general results.

Let G be a group, and let $\varphi \in \mathrm{Hom}(G, \mathrm{End}(G))$. The multiplication on G defined by $x \cdot y = \varphi(x)(y)$ for all $x,y \in G$ induces a ring structure R on G, and in fact all rings R satisfying $R^+ = G$ may be obtained in this manner, 1.2.1.

Notation: Let $\varphi \in \text{Hom}(G, \text{End}(G))$. Then (G,φ) denotes the ring described above. $\ell(A)$ = the left annihilator of A, A a subset of a ring R.

The following observation is easily verified.

Observation 4.6.1: Let $R = (G,\varphi)$ be a ring, $\varphi \in \text{Hom}(G, \text{End}(G))$. Then R is associative if and only if $\varphi[\varphi(x)(y)] = \varphi(x)\cdot\varphi(y)$ for all $x,y \in G$, i.e., φ is a ring homomorphism, $\varphi: R \to \text{End}(G)$.

For the remainder of this section all rings will be assumed to be associative.

Observation 4.6.2: Let $R = (G,\varphi)$, $\varphi \in \text{Hom}(G, \text{End}(G))$. Then $\ell(R) = \ker \varphi$.

Proof: Obvious.

Lemma 4.6.3: Let R be a ring with trivial left annihilator. Let $A,B \subseteq R$ satisfying $R = A+B = \{a+b \mid a \in A, b \in B\}$. If $a_1a_2 = b_1b_2 = 0$ for all $a_i \in A$, $b_i \in B$, $i = 1,2$, then $R = 0$.

Proof: Clearly $A \subseteq \ell(A)$, $B \subseteq \ell(B)$, and $\ell(A) \cap \ell(B) = 0$. Therefore $R^+ = \ell(A)^+ + \ell(B)^+$, and so $\ell(A) = A$, and $\ell(B) = B$. Since $\ell(A)$ is left ideal in R so is A. Let $b \in \ell(A)$, $x \in R$. Then for every $a \in A$, $(bx)a = b(xa) = 0$, i.e., $\ell(A) = A$ is a two-sided ideal in R. Similarly, B is a two-sided ideal in R. Hence $R = A \oplus B$ is a ring direct sum. Clearly $R^2 = 0$ and so $R \subseteq \ell(R) = 0$.

Theorem 4.6.4: Let G be an associative strongly trivial left annihilator group. Then G is indecomposable.

Proof: Suppose that $G = H \oplus K$, $H \neq 0$, $K \neq 0$. Let π_H, π_K be the natural projections of G onto H and K respectively, and let R be a ring satisfying $R^+ = G$, and $R^2 \neq 0$. For $a,b \in R$, define $a\times_H b = \pi_H(a)\cdot\pi_H(b)$, the product on the right being multiplication in R. Then $S = (G, \times_H)$ is a ring with $S^+ = G$, and $K \subseteq \ell(S)$. Therefore $S^2 = 0$ which implies that in the ring R, $H^2 = 0$. Similarly, $K^2 = 0$. By Lemma 4.6.3, $R = 0$, a contradiction.

Corollary 4.6.5: Let G be a non-torsion free group. The following are equivalent:

(1) G is an associative strongly semisimple group.

(2) G is an associative strongly trivial left annihilator group.
(3) $G = Z(p)$, p a prime.

Proof: Clearly (1) $\Rightarrow$ (2). If G satisfies (2) then by Theorem 4.6.4 and Corollary 1.1.5, $G = Z(p^k)$, p a prime, k a positive integer. Let $G = (a)$, $|a| = p^k$. The product $a \cdot a = pa$ induces a ring structure R on G with $p^{k-1}G \subseteq \ell(R)$. Hence $pa = 0$, and $k = 1$. The implication (3) $\Rightarrow$ (1) is obvious.

It is easy to show that if G and H are quasi-isomorphic torsion free groups then G is an (associative) strongly trivial left annihilator group if and only if H is. This together with Theorem 4.6.4 yields:

Corollary 4.6.6: Let G be a torsion free associative strongly trivial left annihilator group. Then G is strongly indecomposable.

Let G be a finite rank torsion free group. The following are some properties of $\dot{E}nd(G)$ which we will need.

Property 4.6.7: $\dot{E}nd(G)$ is a left and right Artinian ring.

Proof: [36, vol. 2, p. 149(d)].

Property 4.6.8: For every idempotent $e \in \dot{E}nd(G)$, $G \doteq e(G) \oplus (1-e)(G)$.

Proof: [36, vol. 2, p. 149 (e)].

Property 4.6.9: $\dot{E}nd(G)$ is a local ring if and only if G is strongly indecomposable.

Proof: [36, Proposition 92.3].

Corollary 4.6.10: Let G be a finite rank, strongly indecomposable torsion free group. Then $\dot{E}nd(G)$ is a local ring whose ideal of non-units is nilpotent.

Proof: $\dot{E}nd(G)$ is a local ring, Property 4.6.9. Let M be the ideal of non-units in End(G). By Property 4.6.8, M does not possess a nonzero idempotent. A right ideal in a right Artinian ring which does not contain a nonzero idempotent is nilpotent, [45, Theorem 1.3.2]. Hence M is nilpotent by Property 4.6.7.

Theorem 4.6.11: Let $R = (G,\varphi)$ G a finite rank, strongly indecomposable torsion free group, and $\varphi \in \text{Hom}(G, \text{End}(G))$. Then either $\ell(R) = 0$, or $\varphi(G)$ is a nilpotent subring of $\text{End}(G)$.

Proof: If $\varphi(G)$ is not nilpotent then by Corollary 4.6.10, $\varphi(G)$ possesses a nonzero unit $\rho \in \dot{\text{End}}(G)$. Since $\dot{\text{End}}(G)$ is a finite dimensional vector space over Q, $\rho^{-1} \in Q[\rho]$, and for some positive integer m, $m\rho^{-1} \in \varphi(G)$. Therefore $m = (m\rho^{-1})\rho \in \varphi(G)$. Put $K = \ker \varphi$. Let $g \in G$ such that $\varphi(g) = m$. Define $\Psi: \varphi(G) \to G$ via $\Psi[\varphi(x)] = \varphi(x)(g)$. Suppose that $\varphi(x) \in \ker \Psi$. Then by Observation 4.6.1, $0 = \varphi[\varphi(x)(g)] = \varphi(x)\varphi(g) = m\varphi(x)$. Since $\varphi(G)^+$ is torsion free, this implies that $\varphi(x) = 0$ and so Ψ is one-to-one. Let $x \in G$ such that $\varphi(x)(g) \in K$. Again by Observation 4.6.1, $0 = \varphi[\varphi(x)(g)] = m\varphi(x)$, implying that $\varphi(x) = 0$, i.e., $\Psi[\varphi(G)] \cap K = 0$. Therefore we have the inclusion $G \supseteq \Psi[\varphi(g)] \oplus K$. For $x \in G$, $\varphi[mx - \Psi(\varphi(x))] = m\varphi(x) - \varphi[\varphi(x)(g)] = m\varphi(x) - \varphi(x)\varphi(g) = m\varphi(x) - m\varphi(x) = 0$. Hence $mx \in \Psi[\varphi(G)] \oplus K$, and so $G \doteq \Psi[\varphi(G)] \oplus K$. Since G is strongly indecomposable, $K = 0$, and R is a trivial left annihilator ring by Observation 4.6.1.

Corollary 4.6.12: Let R be a ring with R^+ a finite rank strongly indecomposable torsion free group, $r(R^+) = n$. Then either $\ell(R) = 0$ or $R^{n+1} = 0$.

Proof: $R = (R^+, \varphi)$, $\varphi \in \text{Hom}(G, \text{End}(G))$. If $\ell(R) \neq 0$ then $\varphi(R^+)$ is nilpotent, Theorem 4.5.11. This clearly implies that R is nilpotent, and so $R^{n+1} = 0$ by Theorem 3.1.3.

Theorem 4.6.13: Let $R = (G,\varphi)$ be a ring with unity, $\varphi \in \text{Hom}(G, \text{End}(G))$. Then $\varphi(G)$ is a direct summand of $\text{End}(G)$.

Proof: $\text{End}(G)$ is a left $\varphi(G)$-module, and $\varphi(G)$ is a projective $\varphi(G)$-module. It therefore suffices to construct a $\varphi(G)$-epimorphism $f:\text{End}(G) \to \varphi(G)$. Define $f(\psi) = \varphi[\psi(e)]$, e the unity of R, for every $\psi \in \text{End}(G)$. Let $x \in G$, $\psi \in \text{End}(G)$. Then $f[\varphi(x)\psi] = \varphi[\varphi(x)\psi(e)] = \varphi(x)\cdot\varphi[\psi(e)] = \varphi(x)\cdot f(\psi)$, and so f is a $\varphi(G)$-homomorphism. For $x \in G$, $f(\varphi(x)) = \varphi[\varphi(x)(e)] = \varphi(x)\varphi(e) = \varphi(x)$, i.e., f is an epimorphism.

Definition: Let G be a torsion free group. A subgroup H of G is a full subgroup of G if G/H is a torsion group. This is clearly equivalent to H being an essential subgroup of G.

Lemma 4.6.14: Let G be a torsion free group V its divisible hull. There is a one-to-one correspondence between the pure fully invariant subgroups of G, and the $\dot{E}nd(G)$-submodules of V.

Proof: Let M be an $\dot{E}nd(G)$-submodule of V. Put $H = G \cap M$. Clearly H is a fully invariant subgroup of G. Since M is a pure subgroup of V, H is a pure subgroup of G. Let H^* be the Q-subspace of V generated by H. Since M is a Q-subspace of V, and H is a full subgroup of M, $H^* = M$.

Conversely, if H is a pure fully invariant subgroup of G, then H^* is an $\dot{E}nd(G)$-submodule of V satisfying $G \cap H^* = H$. Hence the correspondence $H \leftrightarrow H^*$ between pure fully invariant subgroups of G, and $\dot{E}nd(G)$-submodules of V is one-to-one and onto.

The above lemma motivates the following:

Definition: A torsion free group G is irreducible if the only pure fully invariant subgroups of G are 0 and G. If $G \doteq H$ for every nonzero fully invariant subgroup H of G, then G is said to be strongly irreducible.

Corollary 4.6.15: Let A be a simple Q-algebra, R a subring of A. If R^+ is a full subgroup of A^+ then R^+ is irreducible.

Proof: Let H be a pure fully invariant subgroup of R^+, and let $a \in A$. There exists a positive integer n such that $na \in R$. The map $R^+ \to R^+$ via $x \to nax$ is an endomorphism. Hence $naH \subseteq H$. Similarly $H(na) \subseteq H$. This together with the purity of H in R^+ yields that H^* is an ideal in A, and so $H^* = 0$, or $H^* = A$. By Lemma 4.6.14, $H = 0$ or $H = R^+$.

Lemma 4.6.16: Let G be a finite rank strongly indecomposable torsion free group. G is irreducible if and only if $\dot{E}nd(G)$ is a division ring with $\dim_Q \dot{E}nd(G) = r(G)$.

Proof: Suppose that G is irreducible, and let V be the divisible hull of G. By Lemma 4.6.14, V is a simple $\dot{E}nd(G)$-module. Therefore by Schur's Lemma $Hom_{\dot{E}nd(G)}(V,V)$ is a division ring. Since $\dot{E}nd(G) \subseteq Hom_{\dot{E}nd(G)}(V,V)$, $\dot{E}nd(G)$ does not possess a non-zero nilpotent ideal, and so by Corollary 4.6.10, $\dot{E}nd(G)$ is a division ring. Since V is a simple $\dot{E}nd(G)$-module, $\dim_Q \dot{E}nd(G) = \dim_Q V = r(V) = r(G)$.

Conversely, suppose that $\dot{E}nd(G)$ is a division ring satisfying $\dim_Q \dot{E}nd(G) = r(G)$. Then $r(G) = r(V) = \dim_Q V = \dim_{\dot{E}nd(G)} V \cdot \dim_Q \dot{E}nd(G) = \dim_{\dot{E}nd(G)} V \cdot r(G)$. Hence $\dim_{\dot{E}nd(G)} V = 1$, and V is a simple $\dot{E}nd(G)$-module. G is irreducible by Lemma 4.6.14.

<u>Notation</u>: Let K be an algebraic number field, J the ring of algebraic integers in K. For P a prime ideal in J, $J_P = \{x/y \mid x,y \in J,\ y \notin P\}$. For Π a set of prime ideals in J, $J_\Pi = \bigcap_{P \in \Pi} J_P$.

<u>Theorem 4.6.17</u>: Let G be a finite rank strongly indecomposable torsion free group. The following are equivalent.

(1) G is associative semisimple.

(2) G is associative strongly semisimple.

(3) G is quasi-isomorphic to the additive group of a subring J_Π of an algebraic number field K satisfying $[K:Q] = r(G)$, Π an infinite set of primes, or $G \simeq Q^+$.

<u>Proof</u>: (1) ⇒ (3): Let $R = (G,\varphi)$ be a semisimple ring, $\varphi \in \mathrm{Hom}(G, \mathrm{End}(G))$ By Observation 4.6.2, φ is one-to-one. Since the nonzero elements in $\varphi(G)$ are units in $\dot{E}nd(G)$, Corollary 4.6.10, $\varphi(G)^* = Q \otimes \varphi(G)$ is a division algebra over Q. Now $\varphi(G)^+$, identified with $[Z \otimes \varphi(G)]^+$, is a full subgroup of $\varphi(G)^{*+}$ and so $\dot{E}nd(G)$ is a division ring. $\dot{E}nd(G)$ is commutative; an easy consequence of a result in the next chapter, Theorem 5.3.11. Hence $\dot{E}nd(G)$ is a field. Put $K = \dot{E}nd(G)$. By Lemma 4.6.16, $[K:Q] = r(G)$.

Let H be a subgroup of G, $H \neq 0$, satisfying $\varphi(G)(H) \subseteq H$. For $x \in G$, $y \in H$, $\varphi(x)\varphi(y) = \varphi[\varphi(x)(y)]$, Observation 4.6.1. Since $\varphi[\varphi(x)(y)] \in \varphi(H)$ this implies that $\varphi(H)$ is a left ideal in $\varphi(G)$. The same argument used in proving Theorem 4.6.11 shows that $\varphi(H)$ possesses a positive integer m. Therefore $m\varphi(G) \subseteq \varphi(H)$, and so $\varphi(G) \doteq \varphi(H)$. Since φ is one-to-one, $G \doteq H$. Clearly $\varphi(G)(H) \subseteq H$ for every fully invariant subgroup H of G, and so $G \doteq H$ for every nonzero fully invariant subgroup H of G, i.e., G is strongly irreducible. For every $x \in G$,

$x \neq 0$, $H = \varphi(G)(x)$ and $K = \text{End}(G)(x)$ are nonzero subgroups of G satisfying $\varphi(G)(H) \subseteq H$ and $\varphi(G)(K) \subseteq K$. Hence $\varphi(G)(x) \doteq G \doteq \text{End}(G)(x)$. This clearly implies that $\varphi(G) \doteq \text{End}(G)$. Since $\varphi(G)$ is a semisimple ring, the inclusion $m\,\text{End}(G) \subseteq \varphi(G)$, m a positive integer, implies that $\text{End}(G)$ is semisimple. Let J be the ring of algebraic integers in $K = \dot{\text{E}}\text{nd}(G)$, and let Π be the set of prime ideals P in J satisfying $P \supseteq \text{End}(G)$. A standard procedure in algebraic number theory is to define a valuation v_P on K for each prime ideal P in J as follows: Define $v_P(0) = 0$. Let p be the unique rational prime belonging to P. For $x \in J$, $x \neq 0$, define $v_P(x) = p^{-k}$, where k is the highest power of P dividing $\langle x \rangle$ = the ideal in J generated by x. Extend v_P multiplicatively to K. Every non-archimedian valuation on K is equivalent to v_P for some prime ideal P in J. Hence the valuation rings in K are $V_P = \{x \in K \mid v_P(x) \leq 1\}$, P a prime ideal in J. It is readily seen that $V_P = J_P$ for every prime ideal P in J. It is well known [2, p. 91] that if S is a subring of K containing the unity of K, then the integral closure of S is the intersection of all the valuation rings in K containing S. Hence the integral closure of $\text{End}(G)$ is $\bigcap_{P \in \Pi} J_P = J_\Pi$. Therefore by [8, Theorem 2.7] $J_\Pi \doteq \text{End}(G)$. Again the semisimplicity of $\text{End}(G)$ implies that J_Π is semisimple. This is possible only if Π is empty or Π is infinite. If Π is empty, then $J_\Pi = K$ and $\text{End}(G) = K$. Since $G \simeq \varphi(G)^+ \doteq K^+$, and G is indecomposable, $K = Q$, and so $G \simeq Q^+$. The relations $G \simeq \varphi(G)^+ \doteq K^+$ clearly imply that $[K:Q] = r(G)$ and so (1) ⇒ (3).

(1) ⇒ (2): Let $S = (G,\psi)$, $\psi \in \text{Hom}(G, \text{End}(G))$ be a ring with $S^2 \neq 0$. Since $\psi(G) \subseteq \text{End}(G)$, and $\text{End}(G)$ does not possess nonzero nilpotent elements, $\ell(S) = 0$, Theorem 4.6.11, and so ψ is one-to-one by Observation 4.6.1. Therefore $\psi(G)^+$ is a full subgroup of $\text{End}(G)^+$. Let $R = (G,\varphi)$ be a semisimple ring, $\varphi \in \text{Hom}(G, \text{End}(G))$. The map $\varphi(G)^+ \to \psi(G)^+$ via $\varphi(x) \to \psi(x)$ is a $\varphi(G) \cap \psi(G)$ homomorphism. Since $[\varphi(G) \cap \psi(G)]^+$ is a full subgroup of $K^+ = \dot{\text{E}}\text{nd}(G)^+$, this map extends to a K-endomorphism of K^+, i.e., $x \to \alpha x$, $\alpha \in K$. Therefore $\psi(G) = \alpha\varphi(G)$, $\alpha \in K$. Let n be a positive integer such that $n\alpha^{-1} \in \varphi(G) \cap \psi(G)$. Then $\psi(G) \supseteq n\alpha^{-1}\psi(G) = n\alpha^{-1}\alpha\varphi(G) = n\varphi(G)$. Similarly, $\varphi(G) \supseteq n\psi(G)$. Hence $\psi(G) \doteq \varphi(G) \doteq \text{End}(G)$. This implies that $\psi(G)$ is semisimple, which in turn yields that S is semisimple.

(2) $\Rightarrow$ (1): Obvious.

(3) $\Rightarrow$ (1): $G \doteq J_{\Pi}^{+}$, and J_{Π} is semisimple. It is easily seen that if G and H are quasi-isomorphic groups, then G is (associative) semisimple if and only if H is.

For an alternate proof of Theorem 4.6.17 see [6, Theorem 5.5].

In the next chapter, the following group theoretic analogue of the Wedderburn Principal Theorem, due to Beaumont and Pierce, will be proved:

Theorem 5.2.1: Let R be a ring such that R^{+} is a finite rank torsion free group. Then $R^{+} \doteq S^{+} \oplus N^{+}$, where N is the maximal nilpotent ideal of R, S is a subring of R such that $S^* = Q \otimes S$ is a semisimple Q-algebra, and $S \oplus N$ is a ring direct sum. This result will be used in what follows.

Definition: A group G is anti-radical if $J(R) \neq R$ for every ring R with $R^{+} = G$, and $R^2 \neq 0$.

Theorem 4.6.18: Let G be a strongly indecomposable finite rank torsion free group. Then G is associative semisimple if and only if G is anti-radical and not nil.

Proof: If G is associative semisimple, then G is not nil, and by Theorem 4.6.17, G is anti-radical.

Conversely, let G be an anti-radical group. Let $R = (G,\varphi)$, $\varphi \in \text{Hom}(G, \text{End}(G))$ be a ring with $R^2 \neq 0$. Since $\varphi(G)$ is not nilpotent, $\ell(R) = 0$, Theorem 4.6.11, and so by Observation 4.6.2, $G \simeq \varphi(G)^{+}$. Since G is strongly indecomposable, the maximal nilpotent ideal of $\varphi(G)$ is zero by the Beaumont-Pierce Theorem, Theorem 5.2.1. This in turn yields that the nonzero elements in $\varphi(G)$ are units in $\dot{\text{E}}\text{nd}(G)$, Corollary 4.6.10. The argument used in proving Theorem 4.6.11 shows that if $I \triangleleft \varphi(G)$, $I \neq 0$, then I possesses a positive integer m. Hence $m\varphi(G) \subseteq I$ and so $\varphi(G)^{+} \doteq I^{+}$. Therefore if $J[\varphi(G)] \neq 0$, then $\varphi(G)^{+} \doteq J[\varphi(G)]^{+}$ and so $G \doteq J[\varphi(G)]^{+}$. It is easily seen that if G and H are quasi-isomorphic torsion free groups, then G is anti-radical if and only if H is. Therefore $G \doteq J[\varphi(G)]^{+}$ contradicts the fact that G is anti-radical. Hence $J[\varphi(G)] = 0$, i.e., $\varphi(G)$ is semisimple, which implies that R is semisimple.

The above theorem was proved in a different manner in [6, Theorem 5.5].

Theorem 4.6.19: Let G be a finite rank strongly indecomposable torsion free group. If G is the additive group of a ring with trivial left annihilator then G is strongly irreducible, associative strongly trivial left annihilator, and $End(G)$ is a subring of an algebraic number field K satisfying $[K:Q] = r(G)$.

Proof: Let $R = (G,\varphi)$, $\varphi \in Hom(G, End(G))$ be a ring with $\ell(R) = 0$. By Observation 4.6.2, $G \simeq \varphi(G)^+$, and the maximal nilpotent ideal of $\varphi(G)$ is zero by the Beaumont-Pierce Theorem. The nonzero elements in $\varphi(G)$ are units in $\dot{E}nd(G)$, Corollarly 4.6.10, which implies that $\varphi(G)^+ \doteq I^+$ for every nonzero ideal I in $\varphi(G)$, see the proof of Theorem 4.6.11. As in the proof of Theorem 4.6.17, $\varphi(H)$ is a nonzero ideal in $\varphi(G)$ for every nonzero fully invariant subgroup H of G. This implies that $\varphi(G) \doteq \varphi(H)$ which in turn yields $G \doteq H$, and so G is strongly irreducible. The same argument used to prove the implication (1) ⇒ (3) in the proof of Theorem 4.6.17, shows that $K = \dot{E}nd(G)$ is a field satisfying $[K:Q] = r(G)$.

Let $S = (G,\psi)$, $\psi \in Hom(G, End(G))$ be a ring satisfying $S^2 \neq 0$. Then $\psi(G)$ is a nonzero subring of $End(G)$ and so is certainly not nilpotent. By Theorem 4.6.11, $\ell(S) = 0$, and so G is an associative strongly trivial left annihilator group.

Corollary 4.6.20: Let G be a finite rank strongly indecomposable torsion free group with $r(G) = n$. Then either G is an associative strongly trivial left annihilator group, or $R^{n+1} = 0$ for every ring R with $R^+ = G$.

Proof: Theorem 4.6.19, and Corollary 4.6.12.

§7: E-rings and T-rings:

In [35, Problem 45], L. Fuchs posed the problem of classifying the rings R satisfying $R \simeq End(R^+)$. Much progress has been made on this question by P. Schultz, [59], and by R.A. Bowshell and P. Schultz [13]. Many of their results are presented here.

Lemma 4.7.1: If $R \simeq End(R^+)$, and $R^+ = H \oplus D$ with H reduced, and D divisible, then either $D = 0$, or $D \simeq Q^+$ and H is a torsion group.

Proof: Let $r_0(D) = \alpha$. Then R^+ has a direct summand isomorphic to $\bigoplus_\alpha Q^+$, and so $R^+ \simeq End(R^+)^+$ has a direct summand isomorphic to $End(\bigoplus_\alpha Q^+) \simeq \Pi_\alpha(\bigoplus_\alpha Q^+)$, [36, vol. 1, p. 183, Example 6]. Therefore D has a direct summand isomorphic to $\Pi_\alpha(\bigoplus_\alpha Q^+)$ and so $r[\Pi_\alpha(\bigoplus_\alpha Q^+)] \leq \alpha$. This clearly implies that $\alpha = 0$ or 1. Suppose that $Z(p^\infty)$ is a direct summand of D for some prime p. Then R^+ has a direct summand isomorphic to $End(Z(p^\infty))^+ \simeq Z_p^+$. However Z_p^+ has a homomorphic image isomorphic to $\bigoplus_c Q^+$, c = the cardinality of the continuum, and so D has a direct summand isomorphic to $Hom(\bigoplus_c Q^+, Z(p^\infty))$ which is an infinite rank divisible torsion free group, a contradiction. Hence $D = 0$, or $D \simeq Q^+$.

Suppose that $D \simeq Q^+$. Then R^+ has a direct summand isomorphic to $Hom(R^+, Q^+) \simeq \Pi_{r_0(R^+)} Q^+$. Hence $r_0(R^+) = 1$, and H is a torsion group.

Lemma 4.7.2: Let R be a ring satisfying $R \simeq End(R^+)$. Then for each prime p, $R_p^+ = Z(p^{k_p})$ with $0 \leq k_p < \infty$.

Proof: Let p be a prime for which $R_p^+ \neq 0$. By Lemma 4.7.1, R_p^+ is reduced. For every positive integer n, let r_n = the number of cyclic direct summands of R_p^+ of order p^n. Let k be the smallest positive integer for which $r_k \neq 0$. Then R_p^+ has a direct summand $B_k = \bigoplus_{r_k} Z(p^k)$. Since B_k is a pure bounded subgroup of R^+, B_k is a direct summand of R^+, [36, Theorem 27.5]. Therefore R^+ has a direct summand isomorphic to $Hom(B_k, B_k) = \Pi_{r_k}(\bigoplus_{r_k} Z(p^k))$, [36, Corollary 43.3]. Hence $r[Hom(B_k, B_k)] = 2^{r_k} \cdot r_k$ if r_k is infinite and r_k^2 if r_k is finite. Since $r[Hom(B_k, B_k)] \leq r_k$, $r_k = 1$. Let j be a positive integer, $j > k$, and suppose $Z(p^j)$ is a direct summand of R_p^+. Then R_p^+ has a direct summand isomorphic to $Hom(Z(p^k), Z(p^k) \oplus Z(p^j)) = Z(p^k) \oplus Z(p^k)$ contradicting the fact that $r_k = 1$. Therefore $R_p^+ = Z(p^k)$.

Lemma 4.7.3: Let R be a ring satisfying $R \simeq End(R^+)$, and let p be a prime such that $R_p^+ \neq 0$. Then R_p^+ has a unique complement $R_p^1 = \{x \in R \mid h_p(x) = \infty\}$, R is a ring direct sum $R = R_p \oplus R_p^1$, and

$R_p^1 \simeq End(R_p^1)$.

Proof: $R_p^+ = Z(p^k)$, $0 < k < \infty$, Lemma 4.7.2, and so $R^+ = Z(p^k) \oplus H$, [36, Theorem 27.5]. If $pH \neq H$, then R^+ has a direct summand isomorphic to $Hom[R_p^+ \oplus (H/pH), R_p^+] = K$. However $r_p(K) > 1 = r_p(R^+)$, a contradiction. Therefore H is p-divisible, and so $H \subseteq R_p^1$. Conversely, let $x \in R_p^1$. Then $x = y+z$, $y \in R_p^+$, $z \in H$. Since $h_p(x) = h_p(z) = \infty$, $h_p(y) = h_p(x-z) = \infty$. This clearly implies that $y = 0$, and so $x \in H$, i.e., $H = R_p^1$. Clearly R_p and R_p^1 are ideals in R. Let $x \in R_p$, $|x| = p^m$, and let $y \in R_p^1$. Then $y = p^m z$, $z \in R$. Hence $xy = x(p^m z) = (p^m x)z = 0$. Therefore $R_p R_p^1 = 0$, and similarly $R_p^1 R_p = 0$. This clearly yields that $R = R_p \oplus R_p^1$ is a ring direct sum. Now $R \simeq End(R^+) \simeq End(R_p^+) \oplus End(R_p^1) \oplus Hom(R_p^+, R_p^1) \oplus Hom(R_p^1, R_p^+)$. It is easily seen that the last two summands are zero. Therefore $R = R_p \oplus R_p^1 \simeq End(R_p^+) \oplus End(R_p^1)$ with $R_p \simeq End(R_p^+)$, and $R_p^1 \simeq End(R_p^1)$.

Lemma 4.7.4: Let R be a ring satisfying $R \simeq End(R^+)$, and let S be the set of primes p for which $R_p \neq 0$. Put $A = \{x \in R \mid h_p(x) = \infty$ for all $p \in S\}$. Then A is an ideal in R, and R/A is isomorphic to a ring T satisfying $\bigoplus_{p\in S} R_p \leq T \leq \prod_{p\in S} R_p$.

Proof: By Lemma 4.7.3, $R = R_p \oplus R_p^1$ for every $p \in S$. Let π_p be the natural projection of R onto R_p. For $x \in R$ let $\pi(x)$ be the element in $\prod_{p\in S} R_p$ with p-component $\pi_p(x)$ for every $p \in S$. This clearly defines a homomorphism $\pi: R \to \prod_{p\in S} R_p$. Clearly $\ker\pi = A$ and so $\operatorname{im}\pi \simeq R/A$. Now $\pi(R_p) = R_p$ for every $p \in S$, and so $\bigoplus_{p\in S} R_p \leq \operatorname{im}\pi \leq \prod_{p\in S} R_p$.

Observation 4.7.5: Let π be defined as above, and let $U = \prod_{p\in S} R_p$. Then $\operatorname{im}\pi^+$ is a p-pure subgroup of U^+ for every $p \in S$.

Proof: Suppose there exists $p \in S$ such that $\operatorname{im}\pi^+$ is not p-pure in U^+. Now $\operatorname{im}\pi = R_p \oplus C_p$. Since $\operatorname{im}\pi^+$ is not p-pure in U^+, C_p is not p-pure in $\bigoplus_{q\in S, q\neq p} R_q^+$, and so $pC_p \neq C_p$. Therefore R^+ has a direct summand isomorphic to $K = Hom(R_p^+ \oplus (C_p/pC_p), R_p)$. However $r_p(K) > 1 = r_p(R^+)$, a contradiction.

Lemma 4.7.6: Let R be a ring satisfying $R \simeq End(R^+)$, with $R^+ = \bigoplus_{i\in I} G_i$, where $r(G_i) = 1$, and G_i is torsion free for every $i \in I$. Then I is finite, $t(G_i) \nleq t(G_j)$ for $i \neq j$, $G_i \simeq End(G_i)^+$, $t(G_i)$ is idempotent for all $i,j \in I$, and $R \simeq \bigoplus_{i\in I} End(G_i)$.

Proof: R^+ has a direct summand isomorphic to $\prod_{i\in I} End(G_i)$. Hence $r(\prod_{i\in I} End(G_i)) \leq r(R^+) = |I|$. This is possible only if I is finite. Now R^+ has a direct summand isomorphic to $\bigoplus_{i\in I} End(G_i)^+ \oplus \bigoplus_{i\neq j} Hom(G_i, G_j)$. However, $r(\bigoplus_{i\in I} End(G_i)^+) = r(R^+) = |I|$. Therefore $R \simeq \bigoplus_{i\in I} End(G_i^+)$, and $Hom(G_i, G_j) = 0$ for all $i,j \in I$, $i \neq j$. By Proposition 1.3.4, $t(G_i) \nleq t(G_j)$ for all $i,j \in I$, $i \neq j$. Since $t[End(G_i)^+] \leq t(G_i)$ we have that $G_i \simeq End(G_i)^+$. Since $End(G_i)$ is not a zeroring, $t(G_i)$ is idempotent for all $i \in I$, Corollary 2.1.3. For $j \in I$, let π_j be the natural projection of R^+ onto G_j. For $i \in I$, and $a \in R$, the maps $G_i \to G_j$ via $x \to \pi_j(ax)$ and $x \to \pi_j(xa)$ are homomorphisms. Since $Hom(G_i, G_j) = 0$ for $i \neq j$, $\pi_j(ax) = \pi_j(xa) = 0$ for all $x \in G_i$, $i \neq j$, i.e., $RG_i \subseteq G_i$, and $G_iR \subseteq G_i$ and so G_i is an ideal in R for all $i \in I$. Therefore $R = \bigoplus_{i\in I} G_i \simeq \bigoplus_{i\in I} End(G_i)$ is a ring direct sum.

Theorem 4.7.7: Let G be a torsion group. There exists a ring with $R^+ = G$ satisfying $R \simeq End(R^+)$ if and only if G is cyclic.

Proof: Let $G = Z(n)$. Then $(Z/nZ)^+ = G$, and $Z/nZ \simeq End(G)$.

Conversely, let G be a torsion group, and let R be a ring with $R^+ = G$ satisfying $R \simeq End(R^+)$. Let S be the set of primes p for which $G_p \neq 0$. By Lemma 4.7.2, $R^+ = \bigoplus_{p\in S} Z(p^{k_p})$. Since the unity $1 \in R$ has nonzero component in R_p for every $p \in S$, S is a finite set, and so R^+ is cyclic.

Theorem 4.7.8: Let G be the direct sum of rank one groups. There exists a ring R with $R^+ = G$, and $R \simeq End(R^+)$ if and only if $G = Z(n) \oplus \bigoplus_{i=1}^{m} G_i$

with $r(G_i) = 1$, $t(G_i)$ idempotent, $t(G_i) \nleq t(G_j)$ for $i \neq j$, and $pG_i = G_i$ for all primes $p|n$; $i,j = 1,\ldots,m$.

Proof: Let $G = Z(n) \oplus \bigoplus_{i=I}^{m} G_i$ satisfying the conditions of the theorem. Let R_i be a unital subring of Q with $R_i^+ \simeq G_i$, $i = 1,\ldots,m$, see 1.4.8, and put $R = (Z/nZ) \oplus \bigoplus_{i=I}^{m} R_i$. Then $R^+ \simeq G$, and $R \simeq \mathrm{End}(R^+)$.

Conversely, let R be a ring with $R^+ = G$ a direct sum of rank one groups, and $R \simeq \mathrm{End}(R^+)$. Now $R^+ = G_t \oplus H$, with $H = \bigoplus_{i \in I} G_i$, G_i a rank one torsion free group for each $i \in I$. Clearly G_t is an ideal in R. Let p be a prime for which $G_p \neq 0$. Then $G_p = Z(p^k)$, $1 < k < \infty$, Lemma 4.7.2, and so $\mathrm{End}(G_p) \oplus \mathrm{Hom}(H, G_p) = Z(p^k) \oplus \mathrm{Hom}(H, G_p)$ is isomorphic to a direct summand of G. Since $\mathrm{Hom}(H, G_p)$ is a p-group, $\mathrm{Hom}(H, G_p) = 0$. Let $a \in R$, and let π_p be the natural projection of G onto G_p. The maps $H \to G_p$ via $x \to \pi_p(ax)$, and $x \to \pi_p(xa)$ are homomorphisms. Therefore $RH \subseteq H$, and $HR \subseteq H$, i.e., H is an ideal in R. Clearly $G_tH = HG_t = 0$. Hence $R = G_t \oplus H$ is a ring direct sum. Therefore $R \simeq G_t \oplus H \simeq \mathrm{End}(G_t) \oplus \mathrm{End}(H)$ which implies that $G_t \simeq \mathrm{End}(G_t)$, and $H \simeq \mathrm{End}(H)$. By Theorem 4.7.7, $G_t \simeq Z(n)$, and by Theorem 4.7.8 it suffices to show that $pG_i = G_i$ for all $p|n$, and $i \in I$. Suppose that $pG_i \neq G_i$ for some $p|n$ and some $i \in I$. Then R^+ has a direct summand isomorphic to $\mathrm{Hom}(G_p \oplus (G_i/pG_i), G_p)$, and so $r_p(R^+) > 1$, contradicting Lemma 4.7.2.

Theorem 4.7.9: Let G be a group which is not reduced. There exists a ring R with $R^+ = G$, and $R \simeq \mathrm{End}(R^+)$ if and only if $G \simeq Q^+ \oplus Z(n)$, n a positive integer.

Proof: If $G = Q^+ \oplus Z(n)$, then $R = Q \oplus (Z/nZ)$ is a ring satisfying $R^+ \simeq G$, and $R \simeq \mathrm{End}(R^+)$.

Conversely, let R be a ring, with R^+ not reduced, satisfying $R \simeq \mathrm{End}(R^+)$. By Lemmas 4.7.1 and 4.7.2, $R^+ \simeq Q^+ \oplus \bigoplus_{p \text{ a prime}} Z(p^{k_p})$, with $0 \leq k_p < \infty$ for every prime p. Let $p_1,\ldots,p_m$ be the set of primes p for which the unity in R has a non-trivial p-component. Then $R^+ \simeq Q^+ \oplus Z(n)$, with $Z(n) = \bigoplus_{i=1}^{m} Z(p_i^{k_{p_i}})$.

Corollary 4.7.10: Let R be a ring with R^+ a finitely generated group. Then $R \simeq \mathrm{End}(R^+)$ if and only if $R \simeq Z$ or $R \simeq Z/nZ$ for some positive integer n.

Proof: Clearly if $R \simeq Z$ or if $R \simeq Z/nZ$ then $R \simeq \mathrm{End}(R^+)$.

Conversely, suppose that $R \simeq \mathrm{End}(R^+)$ and that R^+ is finitely generated. By Theorem 4.7.8, $R^+ = Z(n) \oplus \bigoplus_{i=1}^{m} G_i$, with $r(G_i) = 1$, and $t(G_i) \neq t(G_j)$ for $i \neq j$, $i,j = 1,\ldots,m$. Since G_i is finitely generated, $G_i \simeq Z^+$ for $i = 1,\ldots,m$, and since $t(G_i) \neq t(G_j)$, $R^+ \simeq Z(n) \oplus Z^+$. For every prime p, $pZ^+ \neq Z^+$, so by Theorem 4.7.8 either $R^+ = Z(n)$ or $R^+ \simeq Z^+$. Since R is a ring with unity, either $R \simeq Z$ or $R \simeq Z/nZ$.

Theorem 4.7.11: Let R be a ring with R^+ a mixed group, S the set of primes p for which $R_p^+ \neq 0$, and such that R^+ has no nonzero elements of infinite p-height for every $p \in S$. Then $R \simeq \mathrm{End}(R^+)$ if and only if:

(1) $R_p \simeq Z/p^{k_p}Z$, $0 < k_p < \infty$ for all $p \in S$,

(2) R is a unital subring of $U = \prod_{p \in S} R_p$, and

(3) R^+ is a p-pure subgroup of U^+ for every $p \in S$.

Proof: Suppose that R satisfies (1) - (3). The exact sequence $0 \to R_t \to R \to R/R_t \to 0$ induces the exact sequence $(*)\ 0 \to \mathrm{Hom}(R^+/R_t^+, R^+) \to \mathrm{End}(R^+) \to \mathrm{Hom}(R_t^+, R^+)$. Since R^+ is p-pure in U^+ for every $p \in S$, R^+/R_t^+ is p-pure in U^+/R_t^+ for every $p \in S$. Therefore the fact that U^+/R_t^+ is p-divisible for every $p \in S$ implies that R^+/R_t^+ is p-divisible for every $p \in S$. This together with the fact that R^+ has no nonzero elements of infinite p-height for $p \in S$, yields that $\mathrm{Hom}(R^+/R_t^+, R^+) = 0$. From $(*)$ we now have that the map $\mathrm{End}(R^+) \to \mathrm{Hom}(R_t^+, R^+)$ via $\varphi \to \varphi_t$ = the restriction of φ to R_t^+ is one-to-one. Since $\varphi(R_t^+) \subseteq R_t^+$ for every $\varphi \in \mathrm{End}(R^+)$, this map is an embedding of $\mathrm{End}(R^+)$ into $\mathrm{End}(R_t^+)$. However $\mathrm{End}(R_t^+) \simeq U$, (in [36, Corollary 43.3] it is shown that $\mathrm{End}(R_t^+) \simeq U^+$, and it is easy to verify that the group isomorphism constructed there is in fact a ring isomorphism). Therefore, for every $f \in \mathrm{End}(R^+)$, there exists $a \in U$ such that $f(x) = ax$ for all $x \in R$. Hence the map $R \to \mathrm{End}(R^+)$ via $a \to a_\ell$ = left multiplication by a is an epimorphism. Since $a(1) = a$, this map is a monomorphism, and so $R \simeq \mathrm{End}(R^+)$.

Conversely, let R be a ring with R^+ a mixed group, and $R \simeq \mathrm{End}(R^+)$. Condition (1) follows from Lemma 4.7.2, condition (2) from Lemma 4.7.4, and condition (3) from Observation 4.7.5.

Theorem 4.7.11 yields the following improvement of Lemma 4.7.4:

Corollary 4.7.12: Let R be a ring with R^+ a mixed group, S the set of primes p for which $R_p^+ \neq 0$, and let $A = \{x \in R^+ \mid h_p(x) = \infty$ for all $p \in S\}$. Then (1) A is an ideal in R, (2) R/A is isomorphic to a ring T satisfying $\bigoplus_{p\in S} R_p \leq T \leq \prod_{p\in S} R_p$, (3) $R/A \simeq \mathrm{End}[(R/A)^+]$, and (4) if $A \simeq \mathrm{End}(A^+)$, then $R \simeq A \oplus (R/A)$.

Proof: (1) - (3) are precisely the conclusions of Lemma 4.7.4. Since R/A satisfies the conditions of Theorem 4.7.11, $R/A \simeq \mathrm{End}[(R/A)^+]$. If $A \simeq \mathrm{End}(A^+)$, then A has unity e. The natural embedding $A \to R$ is reversible by the map $R \to A$ via $x \to ex$, and so $R \simeq A \oplus (R/A)$.

Lemma 4.7.13: Let R be a unital ring, R_ℓ = the ring of left multiplications in R. Then R_ℓ is a ring direct summand of $\mathrm{End}(R^+)$.

Proof: It suffices to reverse the natural embedding $R_\ell \to \mathrm{End}(R^+)$. The reversal is $\mathrm{End}(R^+) \to R_\ell$ via $f \to [f(1)]_\ell$, 1 the unity of R.

Definition: A commutative ring R is called an E-ring if $R \simeq \mathrm{End}(R^+)$.

Lemma 4.7.14: Let R be a ring with unity. The following are equivalent:

(1) R is an E-ring.
(2) The map $R \simeq \mathrm{End}(R^+)$ via $x \simeq x_\ell$ is an isomorphism with inverse $f \to f(1)$, 1 the unity of R.
(3) $\mathrm{End}(R^+) = R_\ell$.

Proof: (1) $\Rightarrow$ (2): By Lemma 4.7.13, R_ℓ is a direct summand of $\mathrm{End}(R^+)$. However $R_\ell \simeq R \simeq \mathrm{End}(R^+)$, and so $R^+ \simeq R^+ \oplus K$. Let $f\colon R^+ \oplus K \to R^+$ be an isomorphism, and let π_{R^+} be the natural projection of $R^+ \oplus K$ onto R^+. Then for every $x \in K$, $f \circ \pi_{R^+}(x) = 0$. However $\mathrm{End}(R^+ \oplus K) \simeq R$ is commutative, and so $\pi_{R^+} \circ f(x) = 0$. Since $\pi_{R^+} \circ f(x) = f(x)$, and f is one-to-one, $x = 0$, i.e., $K = 0$, and so $\mathrm{End}(R^+) = R_\ell$. Therefore the map $R \to End(R^+)$ via $x \to x_\ell$ is an isomorphism. Clearly the inverse of this map is $f \to f(1)$.

(2) ⇒ (3): Obvious.

(3) ⇒ (1): $End(R^+) = R_\ell \simeq R$. Let $x, y \in R$, and let y_r = right multiplication by y. Since $y_r \in End(R^+) = R_\ell$, there exists $z \in R$ such that $y_r = z_\ell$. Hence $xy = y_r(x) = z_\ell(x) = zx$. However $y = y_r(1) = z_\ell(1) = z$, and so $xy = yx$, i.e., R is commutative and therefore is an E-ring.

Corollary 4.7.15: Let R be an E-ring, and let $\varphi \in End(R^+)$. Then $\varphi(R)$ is a principal ideal in R.

Proof: Since $\varphi \in End(R^+) = R_\ell$, there exists $a \in R$ such that $\varphi(x) = ax$ for all $x \in R$, i.e., $\varphi(R) = aR$, and so $\varphi(R)$ is a principal ideal in R.

Corollary 4.7.16: Let R be an E-ring. If H is a group direct summand of R^+, then H is a ring direct summand of R, and so R^+ cannot be an infinite direct sum of non-trivial groups.

Proof: H is a principal ideal in R by Corollary 4.7.15. Let π_H be the natural projection of R^+ onto H. By Lemma 4.7.14(2), $H = \pi_H(R^+) = \pi_H(1)R$. Put $e = \pi_H(1)$. Then $H = eR$, and $e^2 = e \cdot \pi_H(1) = \pi_H^2(1) = \pi_H(1) = e$. Therefore e is an idempotent, and so $H = eR$ is a ring direct summand of R.

Corollary 4.7.17: Let R be a ring with unity. Then $End(R^+)$ is commutative if and only if R is an E-ring.

Proof: If R is an E-ring, then $End(R^+) \simeq R$ is commutative.

Conversely, let R be a ring with unity such that $End(R^+)$ is commutative. Then for $f \in End(R^+)$, $x \in R^+$, $f(x) = f \circ x_r(1) = x_r \circ f(1) = [f(1)]_\ell(x)$, and so $f = [f(1)]_\ell$, i.e., $End(R^+) = R_\ell$. By Lemma 4.7.14, R is an E-ring.

Corollary 4.7.18: Let R be an E-ring, and let S be a ring with unity such that $R^+ \simeq S^+$. Then $R \simeq S$.

Proof: $End(S^+) \simeq End(R^+)$ is commutative, so by Corollary 4.6.17, S is an E-ring. Therefore $S \simeq End(S^+) \simeq End(R^+) \simeq R$.

Corollary 4.7.19: Let R and S be unital rings with torsion free additive groups, and with $R^+ \doteq S^+$. If R is an E-ring then so is S.

Proof: Suppose that R is an E-ring. Then $End(R^+)$ is commutative, and so $\dot{E}nd(R^+) \simeq Q \otimes End(R^+)$ is commutative. Now $End(S^+) \subseteq \dot{E}nd(S^+) \simeq \dot{E}nd(R^+)$, and so $End(S^+)$ is commutative. Therefore S is an E-ring by Corollary 4.7.17.

Lemma 4.7.20: Let G be an E-group, and let $\varphi \in End(G)$. Then $\varphi(G)$ is an E-group, and every endomorphism of $\varphi(G)$ can be extended to an endomorphism of G.

Proof: Let R be an E-ring with $R^+ = G$. By Corollary 4.6.15, $\varphi(G) = eR^+$, $e \in R$. The multiplication $(ex)*(ey) = exy$ for all $x,y \in R$, the product on the right being multiplication in R, induces a ring structure S on $\varphi(G)$. Let 1 be the unity in R. Then $e = e.1$ is the unity in S. Clearly S is commutative. Let $f \in End\ \varphi(G) = End(eR^+)$. It is readily seen that $fe_\ell \in End(G)$. Hence for all $x \in R^+$, $f(ex) = fe_\ell(1)\cdot x = f(e)\cdot x$. Since $f(e) \in eR$, $f(e) = ey$ for some $y \in R$, and so $f(ex) = (ey)\cdot x = (ey)*(ex)$. Therefore $f = (ey)_\ell$, $End(S^+) = S_\ell$, and so S is an E-ring by Lemma 4.7.14. Clearly y_ℓ is an endomorphism of G extending f.

Theorem 4.7.21: Let G be a finite rank strongly indecomposable torsion free group, and let R be a unital ring with $R^+ = G$. Then R is an E-ring.

Proof: Clearly R has trivial left annihilator, so by Theorem 4.6.19, $End(R^+)$ is a subring of an algebraic number field. By Corollary 4.7.17, R is an E-ring.

The following result will be proved in section 4 of the next chapter.

Theorem 4.7.22: Let R be an E-ring with R^+ a finite rank torsion free group. Then $R = \bigoplus_{i=1}^{n} R_i$, with R_i a unital subring of an algebraic number field F_i, and R_i^+ a full subgroup of F_i^+, $i = 1,\dots,n$.

A stronger version of Theorem 4.7.22 may be found in [13, Theorem 3.12].

Theorem 4.7.23: Let R be a unital ring. The following are equivalent:

(1) R is an E-ring.

(2) For $f \in End(R^+)$, $f(1) = 0$ implies that $f = 0$.

Proof: (1) ⇒ (2) by Lemma 4.7.14.

(2) ⇒ (1): Let $f \in End(R^+)$. Then $\varphi = f - f(1)_\ell \in End(R^+)$, and $\varphi(1) = 0$. Hence $f = f(1)_\ell$, i.e., $End(R^+) = R_\ell$, and so R is an E-ring by Lemma 4.7.14.

Definition: A unital ring R is a T-ring if the homomorphism $\mu: R \otimes_Z R \to R$ induced by $\mu(a \otimes b) = ab$ for all $a,b \in R$ is an isomorphism.

From now on $\otimes_Z$ will be denoted by $\otimes$.

Theorem 4.7.24: Every T-ring is an E-ring.

Proof: Let R be a T-ring, and let $d: R \to R \otimes R$ be the map $d(x) = 1 \otimes x$ for all $x \in R$. It is readily seen that d is the inverse of μ, and hence is an isomorphism. For $f \in End(R^+)$, define $\hat{f} \in Hom[(R \otimes R)^+, R^+]$ to be the homomorphism induced by the maps $\hat{f}(a \otimes b) = f(a) \cdot b$ for all $a,b \in R$. If $f(1) = 0$, then $\hat{f}d = 0$. Since d is an isomorphism, $\hat{f} = 0$, which clearly implies that $f = 0$. Hence R is an E-ring by Theorem 4.7.23.

Stronger versions of Theorems 4.7.23 and 4.7.24 may be found in [13, Propositions 1.2 and 1.7].

Theorem 4.7.25: Let G be a torsion group. The following are equivalent:

(1) G is a T-ring group.
(2) G is an E-ring group.
(3) G is cyclic.

Proof: (1) ⇒ (2) by Theorem 4.7.24. (2) ⇒ (3) by Theorem 4.7.7.
(3) ⇒ (1): Let $G = Z(n)$. Then Z/nZ is a T-ring, with $(Z/nZ)^+ \simeq G$.

The following obvious results will soon be employed:

Observation 4.7.26: If R and S are T-rings, then so is $R \otimes S$.

Observation 4.7.27: Every quotient ring of a T-ring is a T-ring.

An immediate consequence of Theorem 4.7.25 is:

Corollary 4.7.28: Let R be a ring with R^+ a torsion group. Then R is a T-ring if and only if $R \simeq Z/nZ$ for some positive integer n.

The problem of classifying T-rings R therefore reduces to the case; R^+ is not a torsion group. The description of these rings is as follows:

<u>Theorem 4.7.29</u>: Let R be a unital ring for which R^+ is not a torsion group. Then R is a T-ring if and only if A) R/R_t is isomorphic to a subring of Q, and B) for every prime p for which $R_p^+ \neq 0$, R_p^+ is cyclic, and R^+/R_p^+ is p-divisible.

<u>Proof</u>: Let R be a T-ring. Clearly, Q is a T-ring, and R/R_t is a T-ring by Observation 4.7.27. Hence $Q \otimes (R/R_t)$ is a T-ring, Observation 4.7.26. Now $Q \otimes (R/R_t)$ is an E-ring, Theorem 4.7.24, and so $[Q \otimes (R/R_t)]^+ \simeq Q^+$ by Theorem 4.7.9. Since $R/R_t \simeq Z \otimes (R/R_t) \subseteq Q \otimes (R/R_t)$, it follows that $r[(R/R_t)^+] = 1$, which implies that R/R_t is isomorphic to a subring of Q, Theorem 4.7.22. Since R is an E-ring, R_p^+ is cyclic for every prime p, Lemma 4.7.2. For p a prime satisfying $R_p^+ \neq 0$, R^+/R_p^+ is p-divisible by Lemma 4.7.3, and so R^+/R_t^+ is p-divisible.

Conversely, let R be a unital ring satisfying A) and B). Consider the commutative diagram:

$$\begin{array}{ccccccccc}
0 & \longrightarrow & R \otimes R_t & \longrightarrow & R \otimes R & \longrightarrow & R \otimes (R/R_t) & \longrightarrow & 0 \\
 & & \downarrow{\scriptstyle \mu_1} & & \downarrow{\scriptstyle \mu_2} & & \downarrow{\scriptstyle \mu_3} & & \\
0 & \longrightarrow & R_t & \longrightarrow & R & \longrightarrow & R/R_t & \longrightarrow & 0 ,
\end{array}$$

where μ_i is induced by the maps $\mu_i(a \otimes b) = ab$, $i = 1, 2,3$.

Since R^+/R_t^+ is p-divisible for every prime p for which $R_p \neq 0$, $R \otimes R_t = \bigoplus_{p \text{ prime}} (R_p \otimes R_p)$. It is readily seen that for each prime p, R_p is a ring direct summand of R, and so $R_p \simeq Z/p^{k_p}Z$ for some non-negative integer k_p. Hence $R_p \otimes R_p \simeq R_p$, and in fact, every element in $R_p \otimes R_p$ may be written in the form $1 \otimes a$, with 1 the unity in R_p. Clearly the restriction of μ_1 to $R_p \otimes R_p$ is an isomorphism for every prime p, and so μ_1 is an isomorphism.

Again the fact that R^+/R_t^+ is p-divisible for every prime p with $R_p \neq 0$, yields that $R \otimes (R/R_t) \simeq (R/R_t) \otimes (R/R_t) \simeq R/R_t$, and in fact every element in $(R/R_t) \otimes (R/R_t)$ is of the form $1 \otimes a$, with 1 the unity in

R/R_t. This clearly implies that μ_3 is an isomorphism, which now yields that μ_2 is an isomorphism, or that R is a T-ring.

Question 4.7.30: An immediate consequence of Theorem 4.7.29 is the fact that if R is a T-ring for which R^+ is not a torsion froup, then 1) $r(R^+/R_t^+) = 1$, 2) $t(R^+/R_t^+)$ is idempotent, 3) R_p^+ is cyclic for every prime p, and 4) for every prime p for which $R_p \neq 0$, R^+/R_p^+ is p-divisible. Is every group satisfying 1) - 4) the additive group of a T-ring? A positive answer, together with Theorem 4.7.25, would provide a complete description of the T-ring groups.

5 Torsion free rings

§1. Notation, definitions, and preliminary results:

Most of the material in this chapter is the work of Beaumont and Pierce on torsion free rings, i.e., rings with torsion free additive group, [7], [8], and [53]. In general for π a group property, a ring R is said to be a π-ring if R^+ is a π-group. A subring S of R is said to be a π-subring of R if S^+ is a π-subgroup of R^+.

Very roughly speaking, the idea of this chapter is to embed a torsion free ring R into the Q-algebra $R^* = Q \otimes R$. It will be shown that if R^* satisfies certain properties, then considerable information is obtained concerning R^+.

All rings in this chapter are assumed to be associative.

Let R, S be subrings of a torsion free ring T. Then R and S are quasi-equal, $R \doteq S$, if there exists a positive integer n such that $nR \subseteq S$, and $nS \subseteq R$. Two torsion free rings R, S are quasi-isomorphic, $R \simeq S$, if they are isomorphic to quasi-equal rings.

<u>Notation</u>:

$d(G)$ = the maximal divisible subgroup of a group G.

$Z(R)$ = the center of a ring $R = \{x \in R \mid xy = yx \text{ for all } y \in R\}$.

$R^* = Q \otimes R$, R a torsion free ring. The symbol $\otimes$ will often be omitted in order to simplify notation, and in order to view R as a full subring of R^*. The elements of R^* will be written as $\sum_{i=1}^{k} r_i x_i$, $r_i \in Q$, $x_i \in R$, $i = 1,\ldots,k$.

<u>Definition</u>: A torsion free group G is quotient divisible if G possesses a free subgroup F such that G/F is a divisible torsion group.

Quotient divisible will be abbreviated, q.d.

Definition: R^* is called the algebraic type of a ring R. Two rings, R,S have the same algebra type if $R^* \simeq S^*$. A torsion free group G admits a multiplication of algebra type A, A a Q-algebra, if there exists a ring R such that $R^+ = G$, and $R^* \simeq A$.

Theorem 5.1.1: Quasi-isomorphic torsion free rings have the same algebra type.

Proof: Without loss of generality we may consider the case $R \doteq S$, R,S torsion free rings, i.e., there exists a positive integer n such that $nR \subseteq S$, and $nS \subseteq R$. Then $R^* = Q \otimes R = Q \otimes nR \subseteq S^* = Q \otimes S = Q \otimes nS \subseteq Q \otimes R = R^*$, and so $R^* = S^*$.

An immediate consequence of Theorem 5.1.1 is

Corollary 5.1.2: Let A be a Q-algebra, and let G,H be torsion free groups, $G \doteq H$. Then G admits a multiplication of algebra type A if and only if H does.

Lemma 5.1.3: Let R be a torsion free ring, and let $P = P(x_1,\ldots,x_n)$ be a homogeneous polynomial with coefficients in Z, such that $x_1,\ldots,x_n$ are non-commuting variables. Then R satisfies the polynomial identity P if and only if R^* does.

Proof: Suppose that R satisfies the polynomial identity P. Let $a_1,\ldots,a_n \in R^*$. Since R is a full subring of R^* there exists a positive integer m such that $ma_i \in R$, $i = 1,\ldots,n$. Let $d = \deg P$. Then $m^dP(a_1,\ldots,a_n) = P(ma_1,\ldots,ma_n) = 0$. Since R^* is torsion free, $P(a_1,\ldots,a_n) = 0$ and R^* satisfies P. The converse is obvious.

Corollary 5.1.4: Let G and H be quasi-isomorphic torsion free groups. (1) G is the additive group of a non-zeroring satisfying a homogeneous polynomial identity $P = P(x_1,\ldots,x_n)$ if and only if H is. (2) Every ring with additive group G satisfies P if and only if every ring with additive group H satisfies P.

Proof: Corollary 5.1.2 and Lemma 5.1.3.

Lemma 5.1.5: Let R be a torsion free ring. R is T-nilpotent if and only if R^* is T-nilpotent.

Proof: Suppose that R is T-nilpotent. Let $\{a_i\}_{i=1}^{\infty} \subseteq R^*$. There exists a positive integer m_i such that $m_i a_i \in R$, $i = 1,2,\dots$. Since R is T-nilpotent, there exists a positive integer k such that $(m_1a_1)(m_2a_2)_k \dots (m_ka_k) = 0$. Therefore $(\prod_{i=1}^{k} m_i)(\prod_{i=1}^{k} a_i) = 0$. Since R^* is torsion free $\prod_{i=1}^{k} a_i = 0$, and R^* is T-nilpotent.

Corollary 5.1.6: Let G,H be quasi-isomorphic torsion free groups. (1) G is the additive group of a T-nilpotent ring which is not a zeroring if and only H is. (2) Every ring with additive group G is T-nilpotent if and only if every ring with additive group H is T-nilpotent.

Proof: Corollary 5.1.2 and Lemma 5.1.5.

Lemma 5.1.7: Let R be a torsion free ring, α an arbitrary ordinal. Then R is α-nilpotent if and only if R^* is.

Proof: An easy induction argument shows that $(R^*)^{\beta} = (R^{\beta})^*$ for every ordinal β. Therefore if R is α-nilpotent, then $(R^*)^{\alpha} = (R^{\alpha})^* = 0$. The converse is obvious.

Corollary 5.1.8: Let G and H be quasi-isomorphic torsion free groups, α an ordinal. (1) G is the additive group of an α-nilpotent ring which is not a zeroring if and only if H is. (2) Every ring with additive group G is α-nilpotent if and only if every ring with additive group H is α-nilpotent.

Proof: Corollary 5.1.2 and Lemma 5.1.7.

§2. The Beaumont-Pierce Decomposition Theorem:

The group theoretic analogue of the Wedderburn Principal Theorem, which was used in Chapter 4, section 5, will be proved here. Since the proof involves several steps, it will be stated before the preliminary results leading up to it are given.

Theorem 5.2.1: Let R be a finite rank torsion free ring, $R^* = \overline{S} \oplus \overline{N}$ a Q-space decomposition of R^*, with $\overline{N}$ the nil radical of R^*, and $\overline{S}$ a semisimple subalgebra of R^*. Put $S = R \cap \overline{S}$, $N = R \cap \overline{N}$. Then S is a subring of R satisfying $S^* = \overline{S}$, N is the maximal nilpotent ideal of R, and satisfies $N^* = \overline{N}$. In addition $S \oplus N$ has finite index in R.

Lemma 5.2.2: Let $S_1 = \{x \in \overline{S} \mid \text{there exists } y \in \overline{N} \text{ for which } x+y \in R\}$. Then $R^+/(S \oplus N)^+ \simeq S_1^+/S^+$:

Proof: Let $\pi_{\overline{S}}$ be the natural projection of R^* onto $\overline{S}$. Define $\varphi: R^+ \to S_1^+/S^+$ via $\varphi(z) = \pi_{\overline{S}}(z) + S^+$ for all $z \in R^+$. Clearly φ is an epimorphism, and $(S \oplus N)^+ \subseteq \ker \varphi$. Let $z \in R$, $z = x+y$, $x \in \overline{S}$, $y \in \overline{N}$. Suppose that $z \in \ker \varphi$. Then $x \in S$, and so $y = z - x \in R \cap \overline{N} = N$, i.e., $z \in (S \oplus N)^+$, and so $\ker \varphi = (S \oplus N)^+$, or $R^+/(S + N)^+ \simeq S_1^+/S^+$.

Lemma 5.2.3: $S^* = \overline{S}$, and $N^* = \overline{N}$.

Proof: S^* is the unique minimal Q-algebra containing S. It therefore suffices to show that S is a full subring of $\overline{S}$. Let $x \in \overline{S}$. Then $x \in R^*$ and so there exists a positive integer n such that $nx \in R$. Hence $nx \in R \cap \overline{S} = S$ and so $\overline{S}^+/S^+$ is a torsion group. The same argument shows that $N^* = \overline{N}$.

Lemma 5.2.4: There is a Q-basis $\{x_1,\ldots,x_m\}$ for $\overline{S}$ such that the free subgroup F of $\overline{S}$ generated by this basis is a subring of S.

Proof: Since S^+ is a full subgroup of $\overline{S}$, a maximal independent set $z_1,\ldots,z_m$ in S is a Q-basis for $\overline{S}$. Hence $z_i \cdot z_j = \sum_{k=1}^{m} a_{ijk} z_k$ with $a_{ijk} \in Q$; $i,j,k = 1,\ldots,m$. Let n be a positive integer such that $na_{ijk} \in Z$ for all $i,j,k = 1,\ldots,m$. Then $\{x_i = nz_i,\ i = 1,\ldots,m\}$ satisfies the conditions of the lemma.

Observation 5.2.5: S_1^+/F is a torsion group.

Proof: Clearly S^+/F is a torsion group. It therefore suffices to show that S_1^+/S^+ is a torsion group, or by Lemma 5.2.2 that $R^+/(S \oplus N)^+$ is a torsion group. Let $x \in R$. Then $x \in R^*$, and so $x = y+z$, $y \in \overline{S}$, $z \in \overline{N}$. By Lemma 5.2.3 there exists a positive integer n such that $ny \in S$, and $nz \in N$. Hence $nx = ny + nz \in S \oplus N$, and $R^+/(S \oplus N)^+$ is a torsion group.

Lemma 5.2.6: For all but finitely many primes p, $(S_1^+/F)_p$, and $(S^+/F)_p$

are divisible and equal.

The proof of Lemma 5.2.6 involves many steps, and will be broken down into a series of claims. First we introduce some notation. The degree of nilpotence of $\overline{N}$ will be denoted by t. For p a prime, $k \geq 0$ an integer, $I_k = \{x \in F \mid p^{-k} x \in S\}$, and $J_k = \{x \in F \mid p^{-k} x \in S_1\}$.

Claim 5.2.7: (1) $F = I_0 \supseteq I_1 \supseteq I_2 \supseteq \dots$,

(2) $F = J_0 \supseteq J_1 \supseteq J_2 \supseteq \dots$,

(3) $I_k \subseteq J_k$,

(4) I_k and J_k are two sided ideals in F,

(5) $I_k \cdot I_\ell \subseteq I_{k+\ell}$,

(6) $J_k \cdot J_\ell \subseteq J_{k+\ell}$, and

(7) $J_k^t \subseteq I_k$ for all $k,\ell \geq 0$.

Proof: (1) - (4) are obviously true.

(5): Let $x_1 \in I_k$, $x_2 \in I_\ell$. Then $x_1 = p^k y_1$, and $x_2 = p^\ell y_2$, $y_1, y_2 \in S$. Hence $x_1 x_2 = p^{k+\ell} y_1 y_2$, with $y_1 y_2 \in S$, i.e., $x_1 x_2 \in I_{k+\ell}$.

(6): follows from the same argument used to prove (5).

(7): Let $x \in J_k$. Then $x = p^k y$, $y \in S_1$, and there exists $z \in \overline{N}$ such that $y - z \in R$.

Let $x_1, \dots, x_t \in J_k$, and let $z_1, \dots, z_t \in \overline{N}$ such that $p^{-k} x_i - z_i \in R$, $i = 1, \dots, t$. Suppose, inductively, that $p^{-k} x_{i_1} x_{i_2} \dots x_{i_j} - z_{i_1} z_{i_2} \dots z_{i_j} \in R$ for $1 \leq i_1 < i_2 < \dots < i_j \leq t$, $j < t$. Then

$p^{-k} x_1 x_2 \dots x_j x_{j+1} - z_1 z_2 \dots z_j z_{j+1} = x_1 (p^{-k} x_2 \dots x_{j+1} - z_2 \dots z_{j+1}) +$
$(p^{-k} x_1 - z_1) x_2 \dots x_{j+1} - p^k (p^{-k} x_1 - z_1)(p^{-k} x_2 \dots x_{j+1} - z_2 \dots z_{j+1}) \in R$.

Therefore $p^{-k} x_1 \dots x_t - z_1 \dots z_t \in R$. However $z_1 \dots z_t = 0$, and so $p^{-k} x_1 \dots x_t \in S_1 \cap R \subseteq \overline{S} \cap R = S$. Hence $x_1 \dots x_t \in I_k$, i.e., $J_k^t \subseteq I_k$.

Claim 5.2.8: $I_{k+\ell} \cap p^\ell F = p^\ell I_k$, and $J_{k+\ell} \cap p^\ell F = p^\ell J_k$ for all $k,\ell \geq 0$.

Proof: Let $x \in I_{k+\ell} \cap p^\ell F$. Then $x = p^{k+\ell} y = p^\ell z$, $y \in S$, $z \in F$. Therefore $x = p^\ell (p^k y) = p^\ell z$. Since R^{*+} is torsion free, $p^k y = z$, and so

$z \in I_k$ which yields that $x \in p^\ell I_k$.

Conversely, let $x \in p^\ell I_k$. Then $x = p^\ell y$, $y \in F$, and $y = p^k z$, $z \in S$. Hence $x = p^{k+\ell} z$, and $x \in I_{k+\ell}$. Clearly $x \in p^\ell F$, so $x \in I_{k+\ell} \cap p^\ell F$.

The same argument replacing I by J, and S by S_1 yields that $J_{k+\ell} \cap p^\ell F = p^\ell J_k$.

Notation: $T = S^+/F$, $T_1 = S_1^+/F$.

Claim 5.2.9: (1) $(p^k T_p)[p^\ell] \simeq I_{k+\ell}^+/[p^\ell F \cap I_{k+\ell}]^+$, and
(2) $(p^k T_{1p})[p^\ell] \simeq J_{k+\ell}^+/[p^\ell F \cap J_{k+\ell}]^+$.

Proof: (1) Let $x \in I_{k+\ell}$. Then there exists a unique element $y \in S$ such that $x = p^\ell y$. Define $\varphi: I_{k+\ell} \to T$ via $\varphi(x) = y+F$. It is easily verified that φ is a homomorphism. Since $p^\ell y \in I_{k+1}^+ \subseteq F$, $\varphi(x) \in T_p[p^\ell]$. Now $x = p^{k+\ell} z$, $z \in S$, which implies that $y = p^k z$, and so $\varphi(x) \in (p^k T_p)[p^\ell]$. Let $w \in S^+$ be such that $w + F \in (p^k T_p)[p^\ell]$. Then $w = p^k z$, $z \in S^+$ and $p^\ell w = p^{k+\ell} z \in F$. Let $x = p^{k+\ell} z$. Then $x \in I_{k+\ell}$, and $\varphi(x) = w+F$, i.e., φ is an epimorphism. Now $x \in I_{k+\ell}^+$ belongs to $\ker \varphi$ if and only if $p^{-\ell} x \in F$ which occurs if and only if $x \in p^\ell F$, i.e., $\ker \varphi = [p^\ell F \cap I_{k+\ell}]^+$. Therefore $(p^k T_p)[p^\ell] \simeq I_{k+\ell}^+/[p^\ell F \cap I_{k+\ell}]^+$. A similar argument yields (2).

Claim 5.2.10: Let p be a prime. If F/pF is semisimple then T_p and T_{1p} are divisible and equal.

Proof: Let $\psi: F \to F/pF$ be the canonical ring epimorphism. Claim 5.2.7 and the semisimplicity of F/pF yield:

(A) $\psi(I_k) \subseteq \psi(J_k) = [\psi(J_k)]^t = \psi(J_k^t) \subseteq \psi(I_k)$, and

(B) $\psi(I_k) = [\psi(I_k)]^2 = \psi(I_k^2) \subseteq \psi(I_{2k}) \subseteq \psi(I_k)$, for all $k \geq 0$.

By Claim 5.2.9, $p^k T_p[p] \simeq I_{k+1}^+/[pF \cap I_{k+1}]^+ \simeq \psi[I_{k+1}]^+ = \psi[J_{k+1}]^+ \simeq J_{k+1}^+/[pF \cap J_{k+1}]^+ \simeq (p^k T_1 p)[p]$ for all $k > 0$. Choosing $k = 0$ yields that $T_p[p] \simeq T_{1p}[p]$. Since $T_p[p] \subseteq T_{1p}[p]$, and $T_{1p}[p]$ is finite, this implies that $T_p[p] = T_{1p}[p]$. For every G, $r(G_p) = r(G_p[p])$. Hence $r(T_p) = r(T_{1p})$.

Claim 5.2.9 and (B) yield that $(p^kT_p)[p] \simeq I^+_{k+1}/[pF \cap I_{k+1}]^+ \simeq \psi(I_{k+1})$ $= \psi(I_{2k+2}) \simeq I^+_{2k+2}/[pF \cap I_{2k+2}]^+ \simeq (p^{2k+1}T_p)[p]$ for all $k \geq 0$. This together with the finiteness of $T_p[p]$ yields that $T_p[p] = (pT_p)[p] = (p^3T_p)[p] = \ldots$ This clearly implies that T_p is is divisible. Hence T_p is a divisible subgroup of T_{1p} with $r(T_p) = r(T_{1p}) < \infty$. This implies that $T_p = T_{1p}$.

In order to proceed it is necessary to introduce some facts concerning a finite dimensional algebra A over a field K of characteristic 0. The reader is referred to [14] and [47] for proofs.

Definition: Let $x_1,\ldots,x_n$ be a basis for A, and let $a \in A$. Then $ax_i = \sum_{j=1}^{n} a_{ij}x_j$, $a_{ij} \in K$, $i,j = 1,\ldots,n$. The trace of a, $t(a) = t\,(a_{ij})$ = the trace of the matrix $(a_{ij}) \in M_n(K)$.

Proposition 5.2.11: $t(a)$ is independent of the choice of basis for A.

Definition: Let $X = \{x_1,\ldots,x_n\}$ be a basis for A. The discriminant of A with respect to X is the determinant of the $n{\times}n$ matrix $(t(x_ix_j))$.

Proposition 5.2.12: Let X,Y be two bases for A, and let d_X, d_Y be the discriminant of A with respect to X and Y respectively. Then there exists $\alpha \in F$, $\alpha \neq 0$, such that $d_Y = \alpha^2 d_X$.

Proposition 5.2.12 makes possible the following:

Definition: Let A be a finite dimensional Q-algebra. A has positive discriminant if the discriminant of A with respect to a Q-basis X of A is positive.

Notation: Let $\{x_1,\ldots,x_n\}$ be a basis for A over K, and let $L \supseteq K$ be a field extension. A_L is the set of all expressions $\{\sum_{i=1}^{n} \alpha_i x_i \mid \alpha_i \in L\}$. For $x = \sum_{i=1}^{n} \alpha_i x_i$, $y = \sum_{i=1}^{n} \beta_i x_i$ two arbitrary elements in A_L, define $x+y = \sum_{i=1}^{n} (\alpha_i + \beta_i)x_i$, and $x \cdot y = \sum_{i,j,k=1}^{n} \alpha_i\beta_j c_{ijk}x_k$, where $x_ix_j = \sum_{k=1}^{n} c_{ijk}x_k$, $c_{ijk} \in K$, $i,j,k = 1,\ldots,n$. These operations clearly induce a ring structure on A_L.

Definition: A is separable over K, if A_L is semisimple for every field extension $L \supseteq K$.

Definition: Let A be a Q-algebra with unity e. An order in A is a subring B of A satisfying: 1) $e \in B$, 2) B possesses a basis for A, 3) every element in B is a root of a monic polynomial with coefficients in Z.

The above definition and the following Proposition may be generalized to algebras over arbitrary fields of characteristic 0. We will assume, from now on, that A is a finite dimensional Q-algebra.

Proposition 5.2.13: Every order in A may be embedded in a maximal order.

Proposition 5.2.14: A is separable over Q if and only if A has non-zero discriminant.

Proposition 5.2.15: If A is semisimple then every nonzero ideal in A is a unique product of prime ideals. If $I \subseteq J$ are nonzero ideals in A, and if $mJ \subseteq I$ for some positive integer m, then the same prime factors occur in the prime decomposition of I and J.

Proposition 5.2.16: Let B be a maximal order in A, X a basis for A, $X \subseteq B$, such that the discriminant of A with respect to X is an integer, and let p be a prime. Then B/pB is semisimple if and only if p does not divide the discriminant of A with respect to X.

We are now in a position to prove Lemma 5.2.6 as follows:

Proof of Lemma 5.2.6: By Claim 5.2.10 it suffices to show that F/pF is semisimple for all but finitely many primes. The subring B of $\overline{S}$ generated by F and the unity $e \in \overline{S}$ is an order in $\overline{S}$. By Proposition 5.2.13, there exists a maximal order $\overline{B}$ in $\overline{S}$ such that $\overline{B} \supseteq B$. Since $\overline{S}$ is separable over Q, $\overline{S}$ has non-zero integral discriminant d with respect to a basis contained in $\overline{B}$, Proposition 5.2.14. By Proposition 5.2.16, $\overline{B}/p\overline{B}$ is semisimple for every prime $p \nmid d$. It therefore suffices to show that $F/pF \simeq \overline{B}/p\overline{B}$ for all but finitely many primes p. Let $\theta_p : \overline{B} \to \overline{B}/p\overline{B}$ be the canonical epimorphism, and let θ_p^F be the restriction of θ_p to F. It suffices to show that θ_p^F is onto for all but finitely many primes p. There exists a positive integer n such that $n\overline{B} \subseteq F$, see the proof of Lemma 5.2.4. Let p be a prime such that $p \nmid n$.

Then $\overline{B} + p\overline{B} = n\overline{B} + p\overline{B} \subseteq F + p\overline{B} \subseteq \overline{B} + p\overline{B}$, and so θ_p^F is onto. This concludes the proof of Lemma 5.2.6.

In addition to being a step towards proving Theorem 5.2.1, Lemma 5.2.6 also implies the following:

Corollary 5.2.17: Let G be a finite rank torsion free group. If G admits a multiplication of semisimple algebra type then G is quotient divisible.

Proof: Let R be a ring with $R^+ = G$ such that R^* is semisimple, and let F be the subring of $R = S$ constructed in Lemma 5.2.4. By Lemma 5.2.6, $(R/F)_p^+$ is divisible for all but finitely many primes p. For every prime p, $(R/F)_p^+ = D_p \oplus C_p$, with D_p divisible, and C_p reduced. Since $r[(R/F)_p^+]$ is finite, C_p is finite for every prime p. This together with the fact that $C_p = 0$ for all but finitely many primes p yields that $G/F = D \oplus C$, D a divisible torsion group and C a finite group. Let n be a positive integer such that $nC = 0$. The restriction of the canonical epimorphism $\theta: G \to G/F$ to nG is a homomorphism onto D with kernel nF. Hence nG is q.d. Since $G \simeq nG$, G is q.d.

Observe that the argument at the end of the proof of Corollary 5.2.17 shows that if a torsion free group G possesses a free subgroup F such that G/F is the direct sum of a divisible torsion group, and a bounded group, then G is q.d.

In order to complete the proof of Theorem 5.2.1 it is necessary to investigate the relationship between T_p and T_{1p} for those primes p for which T_p is not divisible. This is done in the following:

Lemma 5.2.18: T_{1p}/T_p is finite for every prime p.

Again several steps will be required to prove the lemma. As above, $\overline{B}$ is a maximal order in $\overline{S}$ containing F and the unity e in $\overline{S}$, and n is a positive integer satisfying $n\overline{B} \subseteq F$. Put $\overline{I}_k = \overline{B}\, I_k \overline{B}$, and $\overline{J}_k = \overline{B}\, J_k \overline{B}$ for all $k \geq 0$.

An immediate consequence of Claim 5.2.7 and the fact that $n\overline{B} \subseteq F$ is the following:

Claim 5.2.19: (1) $n^2\overline{I}_k \subseteq I_k$, (2) $n^2\overline{J}_k \subseteq J_k$, (3) $\overline{I}_{k+1} \subseteq \overline{I}_k$, (4) $\overline{J}_{k+1} \subseteq \overline{J}_k$, (5) $\overline{I}_k \subseteq \overline{J}_k$, (6) $n\overline{I}_k\, \overline{I}_\ell \subseteq \overline{I}_{k+\ell}$, (7) $n\overline{J}_k\, \overline{J}_\ell \subseteq \overline{J}_{k+\ell}$, (8) $n^{t-1}\, \overline{J}_k^t \subseteq \overline{I}_k$,

(9) $n^2(\overline{I}_{k+\ell} \cap p^\ell F) \subseteq p^\ell \overline{I}_k$, for all $k,\ell \geq 0$.

Since $\overline{B}$ is semisimple, every nonzero ideal in $\overline{B}$ is a product of prime ideals. Let $P_1,\dots,P_u$ be the prime ideals in $\overline{B}$ which are factors of the ideals $p\overline{B}$, $n\overline{B}$, $\overline{I}_1$, or $\overline{J}_1$. The relations $n\,\overline{I}_k^2 \subseteq \overline{I}_{2k}$, and $n\overline{J}_k^2 \subseteq n\overline{J}_{2k}$, which follow immediately from Claim 5.2.19 (6) and (7), together with Proposition 5.2.15 yield that $P_1,\dots,P_u$ are the only prime factors of $\overline{I}_k$, and $\overline{J}_k$ for all $k \geq 1$. Therefore $p\overline{B} = P_1^{\alpha_1} P_2^{\alpha_2} \dots P_u^{\alpha_u}$, $n\overline{B} = P_1^{\beta_1} P_2^{\beta_2} \dots P_u^{\beta_u}$, $\overline{I}_k = P_1^{\gamma_{k1}} P_2^{\gamma_{k2}} \dots P_u^{\gamma_{ku}}$, and $\overline{J}_k = P_1^{\delta_{k1}} P_2^{\delta_{k2}} \dots P_u^{\delta_{ku}}$, $\alpha_i, \beta_i, \gamma_{ki}, \delta_{ki}$ non-negative integers for all $k \geq 1$, $i = 1,\dots,u$.

<u>Claim 5.2.20</u>: (1) $(t-1)\beta_i + t\delta_{ki} \geq \gamma_{ki}$, and (2) $2\beta_i + \max\{\ell\alpha_i, \gamma_{k+\ell,i}\} \geq \ell\alpha_i + \gamma_{ki}$ for all $k \geq 1$, $i = 1,\dots,u$.

<u>Proof</u>: Claim 5.2.19 (8) implies (1), and Claim 5.2.19 (9) implies (2).

<u>Claim 5.2.21</u>: There exists an integer $K(\ell)$ such that for all $k \geq K(\ell)$, $\min\{\gamma_{ki}, \ell\alpha_i\} \leq 2\beta_i + \delta_{ki}$ for all $i = 1,\dots,u$.

<u>Proof</u>: If $\alpha_i = 0$ or if $\gamma_{ki} \leq 2\beta_i$ for all k then the inequality clearly holds. Assume, therefore, that $\alpha_i \neq 0$, and that $\gamma_{k_0 i} > 2\beta_i$ for some k_0. Since the sequence $\{\gamma_{ki}\}_{k=1}^{\infty}$ is non-decreasing, $\gamma_{ki} > 2\beta_i$ for all $k \geq k_0$. Therefore 5.2.20(2) implies that $\gamma_{k+\ell,i} \geq \ell\alpha_i$ for all $k \geq k_0$ and so

(2') $2\beta_i + \gamma_{k+\ell,i} \geq \ell\alpha_i + \gamma_{ki}$ for all $k \geq k_0$. Since $\alpha_i \neq 0$, $\ell\alpha_i \to \infty$ as $\ell \to \infty$, so by (2') $\gamma_{ki} \to \infty$ as $k \to \infty$, and by Claim 5.2.20(1), $\delta_{ki} \to \infty$ as $k \to \infty$. This clearly implies Claim 5.2.21.

An immediate consequence of Claim 5.2.21 is

<u>Claim 5.2.22</u>: For every non-negative integer ℓ, there exists an integer $K(\ell)$ such that for $k \geq K(\ell)$, $\overline{I}_k + p^\ell\overline{B} \supseteq n^2\overline{J}_k$.

<u>Proof of Lemma 5.2.18</u>: Let p be a prime. Since $r(T_p) \leq r(T_{1p}) < \infty$ it suffices to show that $d(T_p) = d(T_{1p})$, or since $T_p \subseteq T_{1p}$, that $d(T_{1p}) \subseteq d(T_p)$. By Claim 5.2.21 there exists an integer $K(\ell)$ such that

for $k \geq K(\ell)$, $n^4J_k \subseteq n^4\overline{J}_k \subseteq n^2\ \overline{I}_k + p^\ell n^2\overline{B}$. However $n^2\ \overline{I}_k \subseteq I_k$, Claim 5.2.19(1), and $n\overline{B} \subseteq F$. Hence (A) $n^4J_k \subseteq I_k + p^\ell F$ for all $k \geq K(\ell)$. Let $\theta_\ell: F \to F/p^\ell F$ be the canonical epimorphism. By (A) $\theta_\ell(n^4J_k^+) \subseteq \theta_\ell(I_k^+)$ for all $k \geq K(\ell)$. Now $\theta_\ell(J_k^+) = (J_k^+ + p^\ell F)/p^\ell F \simeq J_k^+/(p^\ell F \cap J_k)^+ \simeq (p^{k-\ell}T_{1p})[p^\ell]$ by Claim 5.2.9(2). Similarly, employing Claim 5.2.9(1), $\theta_\ell(I_k^+) \simeq (p^{k-\ell}T_p)[p^\ell]$. Let $n^4 = p^j m$, m a positive integer such that $(p,m) = 1$. Then $n^4(p^{k-\ell}T_{1p})[p^\ell] = (p^{k+j-\ell}T_{1p})[p^{\ell-j}]$ is isomorphic to a subgroup of $(p^{k-\ell}T_p)[p^\ell]$. Put $\ell = j+1$ and let $k \geq \max\{K(\ell), \ell\}$. Then $(p^{k-1}T_{1p})[p]$ is isomorphic to a subgroup of $(p^{k-\ell}T_p)[p^\ell]$, and so $r(p^{k-1}T_{1p}) \leq r(p^{k-\ell}T_p)$. For k sufficiently large $p^{k-1}T_{1p} = d(T_{1p})$, and $p^{k-\ell}T_p = d(T_p)$. Hence $d(T_{1p}) \subseteq d(T_p)$ and the claim is proved.

Finally, we can prove Theorem 5.2.1.

Proof of Theorem 5.2.1: All that remains to be shown is that $S \oplus N$ is of finite index in R, Lemma 5.2.3. By Lemma 5.2.2 it suffices to show that S_1^+/S^+ is finite. $S_1^+/S^+ \simeq (S_1^+/F)/(S^+/F) = T_1/T \simeq \bigoplus_{p \text{ a prime}} T_{1p}/T_p$, Observation 5.2.5 and Proposition 1.1.1. Now T_{1p}/T_p is finite for every prime p, Lemma 5.2.18, and 0 for all but finitely primes, Lemma 5.2.6. Hence S_1^+/S^+ is finite; thus concluding the proof of Theorem 5.2.1.

Translating Theorem 5.2.1 to groups, we have:

Corollary 5.2.23: Let G be a finite rank torsion free group. Then $G \doteq H \oplus K$, where H admits a multiplication of semisimple type, and K is the additive group of a nilpotent ring N satisfying $N^2 \neq 0$, unless $K = 0$, or G is nil.

§3. Torsion free rings with semisimple algebra type.

The results of the previous section show that, up to quasi-isomorphism, the classification of the additive groups of finite rank torsion free rings reduces to two cases; the additive groups of rings with semisimple algebra type, and the additive groups of nilpotent rings, which are not zerorings. In this section a further reduction will be obtained by showing that a finite rank torsion free group admitting a multiplication of semisimple algebra type is quasi-isomorphic to a direct sum of groups admitting multiplications of simple algebra type. It will further be shown that a finite rank torsion free group admitting a multiplication of simple algebra type is quasi-isomorphic to a direct sum of groups admitting multiplications of field type. Therefore, upto quasi-isomorphism, the classification of the additive groups of finite rank torsion free rings with semisimple algebra type is settled by determining the additive groups of full subrings of algebraic number fields.

Theorem 5.3.1: Let R be a finite rank torsion free ring with semisimple algebra type. Then R contains a subring $S = S_1 \oplus \ldots \oplus S_m$ such that S_i has simple algebra type, $i = 1,\ldots,m$, and R^+/S^+ is finite.

To prove Theorem 5.3.1 we first need:

Lemma 5.3.2: Let R be a torsion free ring with $R^* = A_1 \oplus \ldots \oplus A_m$ a Q-algebra with unity e. Let $R_i = R \cap A_i$. Then $R_i^* = A_i$, $i = 1,\ldots,m$, and $R^+/(R_1 \oplus \ldots \oplus R_m)^+$ is bounded.

Proof: For each $i = 1,\ldots,m$, $A_i^+/R_i^+ = A_i^+/(R \cap A_i)^+ \simeq (A_i^+ + R^+)/R^+ \subseteq R^{*+}/R^+$ which is a torsion group. Therefore for $x \in A_i$, there exists a positive integer n such that $nx \in R_i$, or $x \in \frac{1}{n} R_i$. Hence $A_i \subseteq QR_i = R_i^*$. However R_i^* is the minimal Q-algebra containing R_i, and so $R_i^* = A_i$. Let $e = e_1 + \ldots + e_m$, $e_i \in A_i$, $i = 1,\ldots,m$. Clearly e_i is the unity in A_i, $i = 1,\ldots,m$, and $\{e_1,\ldots,e_m\}$ is a set of mutually orthogonal idempotents in R^*. Let n be a positive integer such that $ne_i \in R$ for $i = 1,\ldots,m$, and let $x \in R$. Since $x \in R^*$, $x = x_1 + \ldots + x_m$, $x_i \in A_i$, $i = 1,\ldots,m$. For each $i = 1,\ldots,m$, $ne_i x_i = ne_i x$ and so $ne_i x_i \in R \cap A_i = R_i$. Therefore $nx = nex = \sum_{i=1}^{m} ne_i x_i \in \bigoplus_{i=1}^{m} R_i$, and so $R^+/\bigoplus_{i=1}^{m} R_i^+$ is bounded.

Proof of Theorem 5.3.1: Since R^* is semisimple, $R^* = \bigoplus_{i=1}^{m} A_i$, with A_i a simple Q-algebra, $i = 1,\ldots,m$. Put $S_i = R \cap A_i$, $i = 1,\ldots,m$. As in Lemma 5.3.2, $S_i^* = A_i$ and so S_i has simple algebra type, $i = 1,\ldots,m$. Let $S = \bigoplus_{i=1}^{m} S_i$. Then R^+/S^+ is a finite rank bounded group, Lemma 5.3.2, and hence finite.

Corollary 5.3.3: Let G be a finite rank torsion free group. Then G admits a multiplication of semisimple algebra type if and only if G is quasi-isomorphic to a direct sum of groups admitting multiplications of simple algebra type.

Proof: If G admits a multiplication of semisimple algebra type, then it follows immediately from Theorem 5.3.1 that G is quasi-isomorphic to a direct sum of groups admitting multiplications of simple algebra type.

Conversely let $G \doteq \bigoplus_{i=1}^{m} G_i$, and let R_i be a ring satisfying $R_i^+ = G_i$, and R_i^* is a simple Q-algebra, $i = 1,\ldots,m$. Put $R = \bigoplus_{i=1}^{m} R_i$. Then $R^* = \bigoplus_{i=1}^{m} R_i^*$ is semisimple.

Definition: Let R be a finite rank torsion free ring with simple algebra type. A subfield F of $Z(R^*)$ is a field of definition of R if there exists an F-basis $x_1,\ldots,x_k$ of R^* in R such that $S = \bigoplus_{i=1}^{k} (R \cap F)x_i$ has finite index in R.

From now on R will be assumed to be a finite rank torsion free ring, R^* will be denoted by A, and F will signify a subfield of $Z(A)$. Most of the remaining results in this section appear in a more general setting in [53].

Lemma 5.3.4: F is a field of definition of R if and only if $\mathrm{Hom}_F(A^+, A^+) \subseteq \dot{\mathrm{E}}\mathrm{nd}(R^+)$.

Proof: Let $x_1,\ldots,x_n$ be an F-basis of A in R. Put $S = \bigoplus_{i=1}^{k} (R \cap F)x_i$.

Suppose that F is a field of definition of R. Then there exists a positive integer n such that $nR \subseteq S$. Let $\varphi \in \mathrm{Hom}_F(A^+, A^+)$. Since R is a full subring of A, there exists a positive integer m such that $m\varphi(x_i) \in R$, $i = 1,\ldots,k$. Now $mn\varphi(R) \subseteq m\,\varphi(S) = \bigoplus_{i=1}^{k} (R \cap F)m\,\varphi(x_i) \subseteq R$, and so $\varphi \in \dot{\mathrm{E}}\mathrm{nd}(R^+)$.

Conversely, suppose that $\mathrm{Hom}_F(A^+, A^+) \subseteq \dot{\mathrm{E}}\mathrm{nd}(R^+)$. Define $\pi_i\colon A \to F$ via $\pi_i(\sum_{j=1}^{k} f_j x_j) = f_i$; $f_j \in F$, $i,j = 1,\ldots,k$. Clearly $\pi_i \in \mathrm{Hom}_F(A^+, A^+)$, and so there exists a positive integer n such that

$n\pi_i(R) \subseteq R$, $i = 1,\ldots,k$. Let $x \in R$. Then $x = \sum_{i=1}^{k} \pi_i(x)x_i$, and $nx = \sum_{i=1}^{k} n\pi_i(x)x_i \in S$, i.e., $nR \subseteq S$. Hence R^+/S^+ is a finitely generated bounded group. Therefore R^+/S^+ is finite, and F is a field of definition of R.

Lemma 5.3.5: Let A be a simple Q-algebra, A_ℓ, A_r the subrings of $\mathrm{Hom}_Q(A^+, A^+)$ consisting of left and right multiplications, respectively, by elements of A. Then for every subring E of $\mathrm{Hom}_Q(A^+, A^+)$ containing both A_ℓ and A_r, there exists a subfield $F \subseteq Z(A)$ such that $E = \mathrm{Hom}_F(A^+, A^+)$.

Proof: The fact that E is a simple Q-algebra, and that $E \supseteq A_r, A_\ell$ implies that A^+ is an irreducible E-module. Put $F = \{\varphi \in \mathrm{Hom}_Q(A^+,A^+) | \varphi\psi = \psi\varphi$ for all $\psi \in E\}$. Since the elements of F commute with the elements of A_ℓ and A_r, F is a subfield of $Z(A)$. The Jacobson Density Theorem, and the finite dimensionality of A over Q yield that $E = \mathrm{Hom}_F(A^+, A^+)$.

Theorem 5.3.6: Let A be a simple Q-algebra. Then $F = \{\varphi \in \mathrm{Hom}_Q(A^+, A^+) | \varphi\psi = \psi\varphi$ for all $\psi \in \mathrm{End}(R^+)\}$ is the unique smallest field of definition of R, and $\mathrm{Hom}_F(A^+, A^+) = \dot{\mathrm{E}}\mathrm{nd}(R^+)$.

Proof: Since $\dot{\mathrm{E}}\mathrm{nd}(R^+)$ is a subring of $\mathrm{Hom}_Q(A^+, A^+)$, and $\dot{\mathrm{E}}\mathrm{nd}(R^+) \supseteq A_\ell, A_r$, $\dot{\mathrm{E}}\mathrm{nd}(R^+) = \mathrm{Hom}_F(A^+, A^+)$, Lemma 5.3.5, and proof of Lemma 5.3.5. By Lemma 5.3.4, F is a field of definition of R. Let F' be an arbitrary field of definition of R. Again by Lemma 5.3.4, $\mathrm{Hom}_{F'}(A^+,A^+) \subseteq \dot{\mathrm{E}}\mathrm{nd}(R^+) = \mathrm{Hom}_F(A^+, A^+)$. Therefore $F' = \{\varphi \in \mathrm{Hom}_Q(A^+, A^+) | \varphi\psi = \psi\varphi$ for all $\psi \in \mathrm{Hom}_{F'}(A^+, A^+)\} \supseteq \{\varphi \in \mathrm{Hom}_Q(A^+, A^+) | \varphi\psi = \psi\varphi$ for all $\psi \in \mathrm{Hom}_F(A^+, A^+)\} = F$.

In what follows, A will be assumed to be a simple Q-algebra, F the smallest field of definition of R, and $\dim_F(A) = k$.

Corollary 5.3.7: $Z[\dot{\mathrm{E}}\mathrm{nd}(F^+)] = F$. If $e \in R$, e the unity in A, then $Z[\dot{\mathrm{E}}\mathrm{nd}(R^+)] = R \cap F$.

Proof: $F = Z[\mathrm{Hom}_F(A^+, A^+)] = Z[\dot{\mathrm{E}}\mathrm{nd}(R^+)]$ by Theorem 5.3.6.

Suppose that $e \in R$, and let $\alpha \in Z[\mathrm{End}(R^+)]$. For $\varphi \in \dot{\mathrm{E}}\mathrm{nd}(R^+)$ there exists a positive integer n such that $n\varphi \in \mathrm{End}(R^+)$. Since

$\alpha \in Z[\dot{E}nd(R^+)]$, $\alpha(n\,\varphi) = (n\,\varphi)\alpha$. This clearly implies that $n(\alpha\varphi) = n(\varphi\alpha)$. Since $\dot{E}nd(R^+)$ is torsion free, $\alpha\varphi = \varphi\alpha$ and so $\alpha \in Z[\dot{E}nd(R^+)] = F$. However $\alpha = \alpha(e) \in R$, and so $\alpha \in R \cap F$. Clearly $R \cap F \subseteq Z[End(R^+)]$.

Corollary 5.3.8: Suppose that $A = E$, and that $e \in R$, e the unity of A. Then $R = End(R^+)$.

Proof: By Theorem 5.3.6, $End(R^+) = Hom_F(F^+, F^+) = F$, and so $End(R^+)$ is commutative. Therefore by Corollary 5.3.7, $End(R^+) = Z[End(R^+)] = R \cap F = R$.

Corollary 5.3.9: Let H, K be subgroups of R^+ such that $H \cap K = 0$. If $R^+ \doteq H \oplus K$, then QH and QK are F-subspaces of A.

Proof: Clearly $A = QH \oplus QK$. Let π_H be the natural projection of A onto QH. There exists a positive integer n such that $nR \subseteq H \oplus K$. Therefore $n\pi_H(R) = \pi_H(nR) \subseteq \pi_H(H \oplus K) = H \subseteq R$, and so $\pi_H \in \dot{E}nd(R^+) = Hom_F(A^+, A^+)$. Hence QH and similarly QK, are F-subspaces of A.

Corollary 5.3.10: Let A be a field. Then $A = F$ if and only if R^+ is strongly indecomposable.

Proof: Let $x_1, \ldots, x_k$ be an F-basis for A in R. Since F is a field of definition for R, $R \doteq \bigoplus_{i=1}^{k} (R \cap F)x_i$. Therefore if R^+ is strongly indecomposable, $k = 1$, and $A = F$.

Conversely, if R^+ is not strongly indecomposable, then by Corollary 5.39, $k = \dim_F A \geq 2$, and so $A \neq F$.

Piecing together the results of this section we obtain

Theorem 5.3.11: Let G be a torsion free group of finite rank. Then G admits a multiplication of semisimple type if and only if:

(1) G is quotient divisible, and

(2) $G \doteq \bigoplus_{i=1}^{k} G_i$, with G_i a strongly indecomposable group which admits a multiplication of field type $\dot{E}nd(G_i)$, such that $\dot{E}nd(G_i)$ is an algebraic number field satisfying $[\dot{E}nd(G_i) : Q] = r(G_i)$, $i = 1, \ldots, k$.

Proof: Suppose that G admits a multiplication of semisimple type. Then G is quotient divisible, Corollary 5.2.17. By Corollary 5.3.3 we may assume that G admits a multiplication of simple algebra type. Let R be a ring with $R^+ = G$ and R^* a simple Q-algebra. Let $F \subseteq Z(R^*)$ be the smallest field of definition of R, and let $x_1,\ldots,x_k$ be an F-basis for R^* in R. Then $R \doteq \bigoplus_{i=1}^{k} (R \cap F)x_i$. Put $R_i = (R \cap F)x_i$, and $(R \cap F)x_i^+ = G_i$, $i = 1,\ldots,k$. Then $R_i^* \simeq F$ and so G_i admits a multiplication of type F, a field. By Corollary 5.3.8, $R_i = \mathrm{End}(G_i)$ and so $F = R_i^* = \mathrm{End}(G_i)^* = \dot{\mathrm{End}}(G_i)$, $i = 1,\ldots,k$. The indecomposability of G_i, $i = 1,\ldots,k$ follows from Corollary 5.3.10. Now $[\dot{\mathrm{End}}(G_i): Q] = \dim_Q R_i^* = r(G_i)$, $i = 1,\ldots,k$.

The converse is obvious.

§4. Applications.

The results of the previous sections of this chapter will be applied to solve problems which arose in Chapter 3, and to prove Theorem 4.7.22.

Proof of Theorem 3.2.3: Let $G = \bigoplus_{i=1}^{k} G_i$ be a finite rank torsion free group, and suppose that every ring R_i with $R_i^+ \doteq G_i$ is nilpotent. Suppose there exists a ring R with $R^+ = G$ such that R is not nilpotent. Then by Corollary 5.2.23, $G \doteq H \oplus K$, $H \neq 0$, where H admits a multiplication of semisimple type. By Theorem 5.3.11(2), $G \doteq L \oplus M$, with L a strongly indecomposable group admitting a multiplication of field type. Let $G_i \doteq \bigoplus_{j=1}^{k_i} G_{ij}$, G_{ij} strongly indecomposable, $i = 1,\ldots,k$; $j = 1,\ldots,k_i$. By Jonsson's theorem, [36, Theorem 92.5], $G_{ij} \doteq L$ for some $1 \leq i \leq k$, and for some $1 \leq j \leq k_i$. Now $G_i \doteq G_{ij} \oplus H_{ij}$. By Corollary 5.1.2, G_{ij} admits a multiplication of field type. Let S_i be a ring with field algebra type, $S_i^+ = G_{ij}$, and let T_i be any ring with $T_i^+ = H_{ij}$. Put $R_i = S_i \oplus T_i$. Then $R_i^* = S_i^* \oplus T_i^*$ with S_i^* a field. By Lemma 5.1.3, R_i is not nilpotent, a contradiction, Corollary 5.1.4(2).

In Theorem 3.1.3 it was shown that if G is a torsion free group, $r(G) = n < \infty$, and if R is a nil ring with $R^+ = G$, then $R^{n+1} = 0$. The same conclusion follows if R is assumed to be T-nilpotent or α-nilpotent, α an arbitrary ordinal. (Actually, the conclusion for T-nilpotence is obvious since if R is T-nilpotent, then R is nil).

Theorem 5.4.1: Let G be a torsion free group, $r(G) = n < \infty$. Let R be a ring with $R^+ = G$. If R is T-nilpotent or α-nilpotent for some ordinal α, then $R^{n+1} = 0$.

Proof: Suppose that R is not nilpotent. Then $R \doteq S \oplus N$, where S is a subring or R of semisimple algebra type, and N is the maximal nilpotent subring of R, Theorem 5.2.1. By Theorems 5.3.1 and 5.3.6, $S \doteq \bigoplus_{i=1}^{k} S_i$, with S_i a ring of field algebra type. Hence $R^* = \bigoplus_{i=1}^{k} S_i^* \oplus N^*$, with S_i^* a field, $i = 1,\ldots,n$. Clearly S_i^* is neither T-nilpotent nor α-nilpotent, so by Lemmas 5.1.5 and 5.17, S_i is neither T-nilpotent nor α-nilpotent, $i = 1,\ldots,k$. This clearly implies that R is neither R-nilpotent nor α-nilpotent. Therefore if R is T-nilpotent, or α-nilpotent for some ordinal α, then R is nilpotent, so by Theorem 3.1.3, $R^{n+1} = 0$.

Corollary 5.4.2: Let G be a finite rank torsion free group, $r(G) = n$. The following are equivalent:

(1) $R^{n+1} = 0$ for every ring R with $R^+ = G$.
(2) Every ring R with $R^+ = G$ is nilpotent.
(3) Every ring R with $R^+ = G$ is T-nilpotent.
(4) Every ring R with $R^+ = G$ is nil.
(5) Every ring R with $R^+ = G$ is α-nilpotent for some ordinal α.

Proof: The implications (1) $\Rightarrow$ (2) $\Rightarrow$ (3) $\Rightarrow$ (4) and (1) $\Rightarrow$ (5) are obvious. (4) $\Rightarrow$ (1) by Theorem 3.1.3, and (5) $\Rightarrow$ (1) by Theorem 5.4.1.

Proof of Theorem 4.7.22: By Theorem 5.2.1, $R = S \oplus N$, where S is a ring with semisimple algebra type, and N is a nilpotent ring. Since R is unital, $N = 0$. Therefore $R^* = S^*$ is a commutative semisimple Q-algebra, and so $R^* = \bigoplus_{i=1}^{n} F_i$, with F_i an algebraic number field, $i = 1,\ldots,n$. Put $R_i = R \cap F_i$. Then $R = \bigoplus_{i=1}^{n} R_i$, with R_i a unital subring of F_i, and R_i^+ a full subgroup of F_i^+, $i = 1,\ldots,n$.

Bibliography

1 Arnold, D., Finite Rank Torsion Free Abelian Groups and Subrings of Finite Dimensional Q-Algebras, to appear, Springer-Verlag.

2 Bachman, G., Introduction to p-adic Numbers and Valuation Theory, Academic Press, New York-London, 1964.

3 Baer, R., Abelian groups without elements of finite order, Duke Math. J. 3 (1937), 68-122.

4 Bass, H., Finistic dimension and a homological generalization of semi-primary rings, Trans. Amer. Math. Soc. 95 (1960), 466-488.

5 Beaumont, R.A., Rings with additive group which is the direct sum of cyclic groups, Duke Math. J. 15 (1948), 367-369.

6 Beaumont, R.A., Lawver, D.A., Strongly semisimple abelian groups, Pac. J. Math. 53 (1974), 327-336.

7 Beaumont, R.A., Pierce, R.S., Torsion free rings, Ill., J. Math. 5 (1961) 61-98.

8 Beaumont, R.A., Pierce, R.S., Subrings of algebraic number fields, Acta Sci. Math. Szeged 22 (1961), 202-216.

9 Beaumont, R.A., Pierce, R.S., Torsion free groups of rank two, Mem. Amer. Math. Soc. Nr. 38 (1961).

10 Beaumont, R.A., Wisner, R.J., Rings with additive group which is a torsion free group of rank two, Acta Sci. Math. Szeged 20 (1959), 105-116.

11 Beaumont, R.A., Zuckerman, H.S., A characterization of the subgroups of the additive rationals, Pac. J. Math. 1 (1951), 169-177.

12 Borho, W., Uber die abelschen Gruppen auf denen sich nur endlich viele wesentlich verschiedene Ringe definieren lassen, Abh. Math. Sem. Univ. Hamburg 37 (1972), 98-107.

13 Bowshell, R.A., Schultz, P., Unital rings whose additive endomorphisms commute, Math. Ann. 228 (1977), 197-214.

14 Deuring, M., Algebren Springer-Verlag, Berlin, 1968.

15 Feigelstock, S., On the nilstufe of the direct sum of two groups, Acta Math. Sci. Budapest 22 (1971), 449-452.

16 Feigelstock, S., The nilstufe of the direct sum of rank one torsion free groups, Acta Math. Sci. Budapest 24 (1973), 269-272.

17 Feigelstock, S., The nilstufe of homogeneous groups, Acta Sci. Math. Szeged 36 (1974), 27-28.

18 Feigelstock, S., The absolute annihilator of a group modulo a subgroup, Publ. Math. Debr. 23 (1976), 221-224.

19 Feigelstock, S., Extensions of partial multiplications and polynomial identities on abelian groups, Acta Sci. Math. Szeged 38 (1976), 17-20.

20 Feigelstock, S., On groups satisfying ring properties, Comment. Math. Univ. Sancti Pauli 25 (1976), 81-87.

21 Feigelstock, S., The nilstufe of rank two torsion free groups, Acta Sci. Math. Szeged 36 (1974), 29-32.

22 Feigelstock, S., On the type set of groups and nilpotence, Comment. Math. Univ. Sanct Pauli 25 (1976), 159-165.

23 Feigelstock, S., On the generalized nilstufe of a group, to appear in Publ. Math. Debr.

24 Feigelstock, S., An embedding theorem for weakly regular and fully idempotent rings, Comment. Math. Univ. Sancti Pauli 27 (1978), 101-103.

25 Feigelstock, S., The additive groups of subdirectly irreducible rings, Bull. Aust. Math. Soc. 20 (1979), 164-170.

26 Feigelstock, S., The additive groups of subdirectly irreducible rings II, Bull. Aust. Math. Soc. 22 (1980), 407-409.

27 Feigelstock, S., The additive groups of rings possessing only finitely many ideals, Comment. Math. Univ. Sancti Pauli 28 (1980), 209-213.

28 Feigelstock, S., The additive groups of local rings, Archiv for Mat. 18 (1980), 49-51.

29 Feigelstock, S., A simple proof of a theorem of Stratton, Comment. Math. Univ. Sancti Pauli 29 (1980), 21-23.

30 Feigelstock, S., The additive groups of rings with totally ordered lattice of ideals, Quaestiones Mathematicae 4 (1981), 331-335.

31 Feigelstock, S., On a problem of F.A. Szasz, submitted.

32 Feigelstock, S., Schlussel, Z., Principal ideal and Noetherian groups, Pac. J. Math. 75 (1978), 87-92.

33 Freedman, H., On the additive group of a torsion free ring of rank two, Publ. Math. Debr. 20 (1973), 85-87.

34 Fried, E., On the subgroups of an abelian group that are ideals in every ring, Proc. Colloq. Ab. Groups, Budapest (1964), 51-55.

35 Fuchs, L., Abelian Groups, Akadémiai Kiadó, Budapest (1966).

36 Fuchs, L., Infinite Abelian Groups, 1 (1971), 2 (1973), Academic Press, New York-London.

37 Fuchs, L., Ringe und-ihre additive Gruppe, Publ. Math. Debr. 4 (1956), 488-508.

38 Fuchs, L., On quasi-nil groups, Acta Sci. Math. Szeged 18 (1957), 33-43.

39 Fuchs, L., Halperin, I., On the embedding of a regular ring in a regular ring with identity, Fund. Math. 54 (1964), 285-290.

40 Fuchs, L., Rangaswamy, K.M., On generalized regular rings, Math. Z. 107 (1968), 71-81.

41 Gardner, B.J., Rings on completely decomposable torsion free groups, Comment. Math. Univ. Carol. 15 (1974), 381-392.

42 Gardner, B.J. Some aspects of T-nilpotence, Pac. J. Math. 53 (1974), 117-129.

43 Gardner, B.J., Jackett, D.R., Rings on certain classes of torsion free abelian groups, Comment. Math. Univ. Carol. 17 (1976), 493-506.

44 Haimo, F., Radical and antiradical groups, Rocky Mt. J. Math. 3 (1973), 91-106.

45 Herstein, I.N., Noncommutative Rings, Carus Monogr. no. 15, Math. Ass. of Amer., New York, 1968.

46 Jackett, D.R., The type set of a torsion free abelian group of rank two, J. Aust. Soc. Ser. A 27 (1979), 507-510.

47 Jacobson, N., The Theory of Rings, Math. Surveys no. 2, Amer. Math. Soc., New York, 1943.

48 Jacobson, N., Structure of Rings, Colloq. Publ. vol. 37, Amer. Math. Soc. Providence, R.I., 1968.

49 Kertész, A., Volesungen uber Artinsche Ringe, Akadémiai Kiadó, Budapest (1968).

50 Levitzki, J., Contributions to the theory of nilrings (Hebrew, with English summary), Riveon Lematimatka 7 (1953), 50-70.

51 O'Neill, J.D., Rings whose additive subgroups are subrings, Pac. J. Math. 66 (1976), 509-522.

52 O'Neill, J.D., An uncountable Noetherian ring with free additive group, Proc. Amer. Math. Soc. 66 (1977), 205-207.

53 Pierce, R.S., Subrings of simple algebras, Mich. Math. J. 7 (1960), 241-243.

54 Redei, L., Szele, T., Die Ringe "ersten Ranges", Acta Sci. Math. Szeged 12 (1950), 18-29.

55 Ree, R., Wisner, R.J., A note on torsion free nil groups, Proc. Amer. Math. Soc. 7 (1956), 6-8.

56 Reid, J.D., On the ring of quasi-endomorphisms of a torsion free group, Topics in Abelian Groups, Scott-Foresman, Chicago (1963), 51-68.

57 Reid, J.D., On rings on groups, Pac. J. Math. 53 (1974), 229-237.

58 Schlussel, Z., On additive groups of rings (Hebrew, with English summary), Master's thesis, Bar-Ilan University (1976).

59 Schultz, P., The endomorphism ring of the additive group of a ring, J. Aust. Math. Soc. 15 (1973), 60-69.

60 Stratton, A.E., The type set of torsion free rings of finite rank, Comment. Math. Univ. Sancti Pauli 27 (1978), 199-211.

61 Stratton, A.E., Webb, M.C., The nilpotency of torsion free rings with given type set, Comment. Math. Univ. Carol. 18 (1977), 437-444.

62 Stratton, A.E., Type sets and nilpotent multiplications, Acta Sci. Math. Szeged 41 (1979), 209-213.

63 Szász, F., Uber Ringe mit Minimalbedingung fur Hauptrechtsideale, I, Publ. Math. Debr. 7 (1960), 54-64.

64 Szász, F., Uber Ringe mit Minimalbedingung fur Hauptrechtsideale, II, Acta Math. Sci. Budapest 12 (1961), 417-439.

65 Szász, F., Uber Ringe mit Minimalbedingung fur Hauptrechtsideale, III, Acta Math. Sci. Budapest 14 (1963), 447-461.

66 Szász, F., Radikale Der Ringe, Akadémiai Kiadó, Budapest, 1975.

67 Szele, T., Zur Theorie der Zeroringe, Math. Ann. 121 (1949), 242-246.

68 Szele, T., Gruppentheoretische Beziehungen bei gewissen Ringkonstruktionen, Math. Z. 54 (1951), 168-180.

69 Szele, T., Nilpotent Artinian rings, Publ. Math. Debr. 4 (1955), 71-78.

70 Szele, T., Fuchs, L., On Artinian rings, Acta Sci. Math., Szeged 17 (1956), 30-40.

71 Vinsonhaler, C., Wickless, W.J., Completely decomposable groups which admit only nilpotent multiplications, Pac. J. Math. 53 (1974), 273-280.

72 Webb, M.C., A bound for the nilstufe of a group, Acta Sci. Math. 39 (1977), 185-188.

73 Webb, M.C., Nilpotent torsion free rings and triangles of types, Acta Sci. Math. Szeged 41 (1979), 253-257.

74 Wickless, W.J., Abelian groups which admit only nilpotent multiplications, Pac. J. Math. 40 (1972), 251-259.

75 Zariski, O., Samuel, P., Commutative Algebra, Vol. 1, Van Nostrand, Princeton, 1967.

Index